SPIN QUANTUM DYNAMICS IN MOLECULAR MAGNETS

by

JOHN J. HENDERSON

A dissertation submitted in partial fulfillment of the requirements
for the degree of Doctor of Philosophy
in the Department of Physics
in the College of Sciences
at the University of Central Florida
Orlando, Florida

Summer Term 2009

Major Professor: Enrique del Barco

UMI Number: 3383660

All rights reserved

INFORMATION TO ALL USERS
The quality of this reproduction is dependent upon the quality of the copy submitted.

In the unlikely event that the author did not send a complete manuscript
and there are missing pages, these will be noted. Also, if material had to be removed,
a note will indicate the deletion.

Dissertation Publishing

UMI 3383660
Copyright 2009 by ProQuest LLC.
All rights reserved. This edition of the work is protected against
unauthorized copying under Title 17, United States Code.

ProQuest LLC
789 East Eisenhower Parkway
P.O. Box 1346
Ann Arbor, MI 48106-1346

© 2009 John J. Henderson

ABSTRACT

Molecular magnets are ideal systems to probe the realm that borders quantum and classical physics, as well as to study decoherence phenomena in nanoscale systems. The control of the quantum behavior of these materials and their structural characteristics requires synthesis of new complexes with desirable properties which will allow probing of the fundamental aspects of nanoscale physics and quantum information processing. Of particular interest among the magnetic molecular materials are single-molecule magnets (SMMs) and antiferromagnetic (AFM) molecular wheels in which the spin state of the molecule is known to behave quantum mechanically at low temperatures. In previous experiments the dynamics of the magnetic moment of the molecules is governed by incoherent quantum tunneling. Short decoherence times are mainly due to interactions between molecular magnets within the crystal and interactions of the electronic spin with the nuclear spin of neighboring ions within the molecule. This decoherence problem has imposed a limit to the understanding of the molecular spin dynamics and the sources of decoherence in condensed matter systems. Particularly, intermolecular dipolar interactions within the crystal, which shorten the coherence times in concentrated samples, have stymied progress in this direction. Several recent works have reported a direct measurement of the decoherence time in molecular magnets. This has been done by eliminating the dephasing created by dipolar interactions between neighboring molecules. This has been achieved by a) a dilution of the molecules in a liquid solution to decrease the dipolar interaction by separating the molecules, and b) by polarizing the spin bath by applying a high magnetic field at low temperatures. Unfortunately, both approaches restrict the experimental studies of quantum dynamics. For example, the dilution of molecular magnets in liquid solution causes a dispersion of the molecular spin orientation and anisotropy axes, while the large fields required to polarize the spin bath overcome the anisotropy of the molecular spin. In this thesis I have explored two methods to overcome dipolar interactions in molecular magnets: a) studying the dynamics of molecular magnets in single crystals where the separation between magnetic molecules is obtained by chemical doping or where the high crystalline quality allows observations intrinsic to the quantum mechanical nature of the tunneling of the spin, and b) studying the electronic transport through an individual magnetic molecule which has been carefully placed in a single-electron transistor device. I have used EPR microstrip resonators to measure $Fe_{17}Ga$ molecular wheels within single crystals of Fe_{18} AFM wheels, as well as demonstrating, for the first time in a Single Molecule Magnet, the complete suppression of a Quantum Tunneling of the Magnetization transition forbidden by molecular symmetry.

ACKNOWLEDGMENTS

I would like to thank the members of my dissertation committee Robert Peale and Stephen Hill for access to their labs and countless experimental guidance, and Eduardo Mucciolo for theoretical explanations.

Thanks to David Bradford for all the help in the machine shop.

Special thanks to my lab mates Chris Ramsey, Juan-Carlos Gonzalez-Ponz, Firoze Haque, Hajrah Moin Quddusi, Simranjeet Singh and Asma Amjad.

Very special thanks to my Advisor Enrique del Barco for his support and dedication.

To my family, without which this would have been impossible. To all of my friends, for their support.

TABLE OF CONTENTS

LIST OF FIGURES viii

CHAPTER 1: INTRODUCTION 1

    1.1 Molecular Magnets 5

        1.1.1 Single-Molecule Magnets 5

        1.1.2. Antiferromagnetics Molecular Wheels 7

    1.2 Sources of Decoherence in Single Crystals of Molecular Magnets 8

CHAPTER 2: THE SINGLE CRYSTAL APPROACH 10

    2.1 Magnetic Dilution of Single Crystals 10

    2.2 Development of Microstrip Resonators for High-Frequency EPR Spectroscopy Studies . 13

        2.2.1 Microstrip Line Resonators 13

        2.2.2 Microstrip Cross Resonators 16

    2.3 EPR Studies of Diluted Fe17Ga Wheels 18

    2.4 Observation of Quantum Mechanical Spin Selection Rules in the SMM Mn3 29

CHAPTER 3: THE SINGLE MOLECULE APPROACH 41

    3.1 . Physics of SET Devices 41

    3.2. Development of SET Devices 43

        3.2.1 Single-Electron Transistors 43

        3.2.2. Fabrication of SET Devices 46

        3.2.3. Functiontionalization of SMMs 51

        3.2.4. Deposition of SMMs 51

        3.2.5. Preliminary Results 53

CHAPTER 4: CONCLUSIONS 55

APPENDIX: CLEANROOM RECIPES 57

REFERENCES 72

LIST OF FIGURES

Fig. 1 (a) Energy representation of the anisotropy barrier of Mn_{12}-acetate, showing the energy levels corresponding to 2S+1 opposite spin projections along the easy magnetic axis of the molecule. (b) Graphic representation of single molecule magnet Mn_{12}. (c) Magnetization curve of Mn_{12}-acetate at T = 0.6 K. The jumps are due to resonant quantum tunneling at the resonance fields. The background shows the potential energy wells tilted by the action of a longitudinal field. 6

Fig. 2: (a) Molecular structure of a Fe_{18} (Fe^{3+}, S = 5/2) antiferromagnetic wheel doped with a single Ga^{3+} (S = 0) ion in one of the six most probable positions (circled in blue). This doping leads to spin-frustration and a magnetic ground state S = 5/2 at low temperature. (b) Representation of the magnetic dilution achieved in a single crystal of Fe_{18} molecular wheels by controlling the average separation between magnetic molecules through Ga-doping 11

Fig. 3: Phase relaxation time originated by dipolar coupling with surrounding molecules as a function of the intermolecular separation 12

Fig. 4: Sketches of a microstrip resonator indicating the characteristic geometrical parameters (up) and functionality (down). 14

Fig. 5: Left: Calculated reflection and transmission S-parameters of a transmission microstrip resonator for four different sizes of the transmission gap, g_t = 140, 300, 500μm and ∞ (reflection resonator, undistinguishable from the 500μm resonator response). Center: Color-code plot of the ac current density (in A/m) at resonance in three different resonators; $g_t = \infty$ (a), $g_t = g_c = 140$μm (b) and $g_t = 400$μm (c). Right: Measured response of a microstrip resonator fabricated on a GaAs substrate, with the following parameters: h = 500μm, L = 3mm and w = 500μm. 15

Fig. 6: Measured response of a microstrip resonator fabricated on a GaAs substrate, with the following parameters: h = 500μm, L = 3mm and w = 500μm. 17

Fig. 7: Optical micrograph of a microstrip resonator, A typical Fe18:Ga crystal can be seen mounted on the 0.4-mm-wide microstrip line resonator with vacuum grease. 18

Fig. 8: 9.7 GHz EPR spectra as a function of the longitudinal magnetic field, H_L, recorded at different temperatures on a single crystal of Fe_{18} molecular wheels doped with Ga at a concentration of 0.005% (Ga:Fe), which corresponds to an average separation between magnetic molecules of ~8 nm. The behavior of the EPR peaks with temperature is indicative of a positive uniaxial anisotropy parameter, D > 0. Three peaks, labeled 1/2, 3/2 and 5/2, correspond to transitions between levels with the same $|S_s|$. A third peak, labeled (-1/2,3/2), corresponds to transitions between states of different spin projection. 19

Fig. 9: Energy level diagram calculated by direct diagonalization of Eqn. 1 with magnetic field aligned with the wheel z-axis, the indicated transitions correspond to the peaks observed in Figure 8. 20

Fig. 10: The transverse field EPR spectra recorded at the same frequency and temperatures. The observed $\Delta_{1/2,-1/2}$ and $\Delta_{3/2,-3/2}$ peaks correspond to transitions between the split superposition states associated to the spin projections $|S_z| = 1/2$ and $|S_z| = 3/2$, respectively[19]. The paramagnetic impurity is observed as a weak peak (α) at ~0.33 T. 22

Fig. 11: Energy level diagram calculated by direct diagonalization of the Hamiltonian given in equatio.2 (see text for parameter values) with the magnetic field aligned perpendicular the wheel z axis. All indicated transitions correspond to the peaks observed in Figure 10. 23

Fig. 12: Position of the 9.7 GHz EPR absorption peaks as a function of the angle of application of the transverse field within the plane of the wheel. The experiments have been performed at T=300 mK on a single crystal of Fe18 molecular wheels doped with Ga at a concentration of 0.005% Ga:Fe. 25

Fig. 13: Behavior of the EPR peak corresponding to the $S_z = \pm 5/2$ transition at 9.7 GHz as a function of the Ga:Fe doping concentration (0.005%-0.5%). 26

Fig. 14: Peak's width versus concentration. The peak narrows for low concentrations and saturates at ~37 Oe below 0.01% Ga:Fe concentration. The dephasing times extracted from the peak's width are shown below some of the data. 27

Fig. 15: Calculated average distance between magnetic molecules (Fe17Ga) as a function of the concentration. The calculated dipolar dephasing time associated to the intermolecular distance is provided for some of the data. The saturation of the peak width together with the difference between the calculated and estimated dephasing times indicate that the origin of the peaks broadening is inhomogeneous. 28

Fig. 16: Field derivative of the magnetization curves obtained for a Mn3-Cl single crystal at different temperatures, with the field swept at 1 T/min along the easy axes of the molecules. 32

Fig. 17: Magnetization versus longitudinal magnetic field recorded at 0.3 K in a Mn3-Cl single crystal. The indicated tunnel splitting values are order of magnitude estimates from the change of magnetization at each resonance. 34

Fig. 18: a) The sweep rate dependence of resonance k=1, b) The dM/dH peak at resonance k = 1 upon application of a transverse magnetic field. 35

Fig. 19: The QTM probability increases monotonically at resonances k = 1 and 2 with increasing transverse field. 36

Fig. 20: Energies of the Mn3 mS levels as a function of the longitudinal magnetic field. Symbols highlight level crossings between different spin states, indicating the degeneracies (red circles) and avoided crossings (blue squares) expected for the C3 symmetry of the structures. Inset: field derivatives of the magnetization curves (dM/dHL) measured for a Mn3-Br single crystal at different temperatures, with field sweep rates of 0.05 T/min (data below 1.7 T) and 0.2 T/min (data above 1.7 T). 37

Fig. 21: Ground state tunnel splittings associated with resonances k = 0 (black squares), k = 1 (red circles), k = 2 (green triangles) and k = 3 (blue stars) as a function of the transverse field, HT, with the Jahn-Teller axes aligned along the z-axis (thin lines) and tilted θ = 8.5o away from the z-axis (thick lines). A few symbols per curve have been added to help in their identification. The grey shaded region in the vicinity of zero transverse field represents the strength of the dipolar magnetic field (~250 G) felt by the Mn3 molecules within the single crystal. The horizontal lines within the grey region indicate the order of magnitude of the splitting values attained at resonances k = 1 and k = 2 for a transverse field of 250 G. 39

Fig. 22: The content of these diagrams is discussed in the text. 41

Fig. 23: (A) Schematic representation of a three terminal SET. (B) Energy representation of the electrostatic levels of the molecule with respect to the Fermi energies of the source and drain electrodes. N represents the molecule on its neutral charge state, being N+1 and N-1 its first reduction and oxidation states, respectively. E_c is the charging energy of the molecule. (C) The effect of a bias voltage on the energy configuration of the molecular SET. Note that the first exited state (N+1) of the molecule is now available for conduction, since it lies within the Fermi energies of the electrodes. 44

Fig. 24: (a) Calculated current for an ideal single-state molecular SET as a function of the bias voltage, V_{BIAS}, for different gate voltages, Vg. b) I and c) dI/dV contour-plots of a molecular SET calculated assuming asymmetric barriers and tunnel rates. The dashed color lines in b correspond to the I-V curves shown in a). 45

Fig. 25: A) Design perspective and B) photograph of our SET device. The chip on B is 1´1 cm2 and contains 35 SETs grouped in five sets of six transistors each, as shown in A. The substrate is pure silicon with a one micrometer silicon oxide layer on top. C) AFM image of a zoomed area in the chip where the wire nanoconstriction and the Al gate are shown. D) SEM image of the gold nanowire. 48

Fig. 26: a) IV curves recorded during electromigration-induced wire breaking for three circuit resistances. b) Zero bias resistance (ZBR) as a function of breaking voltage. c) Distribution of breaking currents for a circuit resistance of 235 W. d) Different IV curves observed after wire breaking: Asymmetric smooth curves (SMH), curves with zero bias enhancement of the current (ZBE), curves with steps (STP), and curves with current suppression at low voltages consistent with Coulomb blockade (CB). e) Pie chart showing the statistics of the IV curve types found after the breaking of over 60 nanowires. NC corresponds to curves where no current was observed after breaking..... 48

Fig. 27: Electromigrated gold wire showing a small gap and gold particles that were removed. 50

Fig. 28: a) bare epitaxial gold surface b) 5 minute deposition of $Mn_{12}O_{12}(TBP)_{12}(O_2CMe)_4]H_2O$ time c) 10 minute deposition time d) 60 minute deposition time. 52

Fig. 29: a) IV curves recorded at 4K with a Mn_{12}-3tpc based SET. b) Differential conductance vs. bias voltage for the same gate voltages of B. c) dI/dV contour plot as a function of the bias and gate voltages showing the typical response of a molecular SET. 53

CHAPTER 1: INTRODUCTION

A potential application in quantum computation and in general quantum technologies has lead to a major push by researchers to control and manipulate spin and charge degrees of freedom in nanoscale systems. Condensed-matter systems have been the prototypical candidate for such technologies, with Josephson junction devices [1], semiconductor quantum dots [2], and single-molecule magnets (SMMs) [3] all having been proposed as qubits. Advances have been obtained with both Josephson junctions and quantum dots, where coherent manipulation of the quantum states has been achieved [1,2]. There has been a large effort by many in the field of molecular magnetism in this direction as well.

Molecular magnets [4-15] are ideal systems to probe the realm that borders quantum and classical physics, as well as to study decoherence phenomena in nanoscale systems. The control of the quantum behavior of these materials and their structural characteristics requires synthesis of new complexes with desirable properties which will allow probing of the fundamental aspects of nanoscale physics and quantum information processing. Of particular interest among the magnetic molecular materials are single-molecule magnets (SMMs) and antiferromagnetic (AFM) molecular wheels in which the spin state of the molecule is known to behave quantum mechanically at low temperatures.

In most experiments the dynamics of the magnetic moment of the molecules is governed by incoherent quantum tunneling (i.e. the characteristic time of the experiment is longer than the decoherence time). Short decoherence times are mainly due to interactions between molecular magnets within the crystal and interactions of the electronic spin with the nuclear spin of neighboring ions within the molecule. The coupling between the magnets and their local crystal environment enhances decoherence processes that disturb the quantum states of the system. All of these specifics have made the standard magnetic characterization approaches inappropriate for the study of quantum dynamics in molecular magnets. This problem has imposed a limit to the understanding of the molecular spin dynamics and the sources of decoherence in these systems. Particularly, intermolecular dipolar interactions within a single crystal, which shorten the coherence times in concentrated samples, have stymied progress in this direction. The work of Ardavan *et al.* reported the first direct measurement of the decoherence time of molecular wheels diluted in a liquid solution [12], and has established the fact that in order to study the effect of nuclear spins on the decoherence of the electron spin in molecular magnets dipolar dephasing needs to be reduced. This experiment has also been achieved by Bertaina *et al.*,[13] and Schlegel *et al.*,[14] using dilution of the molecular nanomagnets in solution until obtaining a significant intermolecular separation. Unfortunately, the dilution of molecular magnets in a liquid solution causes a dispersion of the molecular spin orientation and anisotropy axes which causes the molecules within the sample to respond differently to an applied external field. Following an alternative approach Takahashi *et al.*,[15] use a large magnetic field was used to polarize the spin bath in a condensed crystalline sample of Fe_8 SMMs, a method that was previously shown to work successfully to eliminate the dipolar dephasing mechanism in nitrogen impurities in diamond.[16] However, the spin-bath polarization requirement substantially restricts the experimental conditions of the experiment to low temperatures, large magnetic fields, and high frequencies, which imposes limitations for the phenomenology under study.

During my thesis work I have explored two methods of overcoming the limitations due to dipolar dephasing. The first approach called "the single crystal approach" consists of studying the spin dynamics of molecular magnets in magnetically dilute crystalline environment. For this study I have used EPR microstrip resonators designed and fabricated at UCF to investigate the magnetic behavior of $Fe_{17}Ga$ molecular wheels dispersed within a single crystal of Fe_{18} AFM wheels. The doped wheel is created during the chemical synthesis of the crystals. The level of doping is synthetically controlled in order to tune the distance between molecules with a dipole moment. On each wheel, a Ga ($S = 0$) can replace a Fe ($S = 5/2$) so that the spin of one Fe is noncompensated. This leads to a net spin of 5/2 for a single doped wheel. By controlling the amount of Ga during synthesis we can tailor the crystal to have $Fe_{17}Ga$ magnetic wheels spaced a given distance apart. As the distance between frustrated wheels increases the dipolar dephasing time also increases, until a distance is reached where the dipolar dephasing time is greater than the characteristic decoherence of the system.

The results show that the magnetic dilution process works and that the degree of inhomogeneity can be controlled during synthesis. By varying the concentration of Ga in the system from 0.5% to 0.005% we were able to tune the average intermolecular separation of $S = 5/2$ wheels from 2 up to 8 nm. This separation showed that there was a significant decrease in the degree of inhomogeneity of the system as the magnetic wheels got farther away from eachother.

I have also investigated the quantum dynamics of a novel Mn_3 SMM in the solid state form (single crystal). In this case, I have studied the effect of the quantum mechanical spin selection rules that are imposed by molecular symmetry on the quantum tunneling of the magnetization (QTM). Specifically, I have demonstrated, for the first time in an SMM, the complete suppression of a QTM that is forbidden by the symmetry of the molecule. This SMM takes the form of a highly symmetric equilateral triangle. Due to its threefold symmetry you would expect to see resonances involving states differing in spin by a multiple of 3 ($\Delta m_s=3n$, where n is an integer) but in all previous works every sequential resonance (k=0,1,2,3) is seen regardless of the symmetry of the molecule. In our work we see resonances k=0,2, and 3 prominently and the appearance of k=2 resonance is explained by using local tilts of the Jahn-Teller axes of each Mn molecule in the triangle, these tilts are in excellent agreement with XRD data.

The second approach to study the quantum dynamics of molecular magnets, which I will call the "individual molecule approach", is probing the magnetic behavior of individual molecular magnets by means of electrical transport using a Single-Electron Transistor (SET). The magnetic molecules are attached to electro-migrated nanometer spaced metal electrodes (source and drain) and strongly locally gated to form an SET. Studying an individual molecule and how it behaves during electron transport measurements provides a low current level spectroscopic tool to investigate the physical properties of one individual molecule outside of a local crystal environment. By using only one molecule we remove sources of magnetic diplolar dephasing from the experiment so one can probe the quantum properties without interference from the local environment, which is found in a single crystal of molecular magnets.

1.1 Molecular Magnets

1.1.1 Single-Molecule Magnets

SMMs are a class of molecules containing multiple transition-metal ions bridged by organic ligands. These ions are strongly coupled by exchange interaction, often in a ferrimagnetic manner, yielding large magnetic moments per molecule. The large spin, combined with a zero-field splitting, provides an anisotropy barrier to magnetization reversal (see Figure 1a). SMM crystals offer several advantages relative to other magnetic structures. Most importantly, they are monodisperse (all of the molecules in a crystal have the same spin, anisotropy, and atomic structure), and they are weakly interacting. This monodispersity enables the study of behavior that is intrinsic to the magnetic nanostructure and which has been experimentally inaccessible in other classes of magnetic materials. Particularly important is the appearance of quantum tunneling between opposite spin projection levels of the molecule, which leads to step-wise magnetic hysteresis loops and accelerated magnetic relaxation at fields that switch on the quantum tunneling mechanism [7]. Another advantage involves the weak coupling between the molecule's spin levels and the environment.

These systems have several potential applications, including quantum computation and quantum information storage [3]. Moreover, by containing a single bit per molecule, SMMs represent the ultimate classical limit in magnetic information storage. Thus, they are promising candidates for future magnetic data storage media with an improvement in storage density by several orders of magnitude in comparison with current technologies. Understanding the relationship between the quantum properties of these materials and their structural characteristics will require the synthesis of new complexes with desirable properties, which will, in turn, allow in-depth studies on the fundamental aspects of nanoscale physics and quantum information processing.

Figure 1b shows the spatial disposition of the magnetic (Mn) ions in the prototype and most widely studied SMM, Mn$_{12}$-acetate. This molecule has a central core of four Mn^{+4} ions ($S = 3/2$, green) surrounded by a ring of eight Mn^{+3} ions ($S = 2$, yellow). The central ions are ferromagnetically coupled and the central core is antiferromagnetically coupled with the outer ring, giving rise to a net spin $S = 10$ at low temperatures ($T < 30$ K).

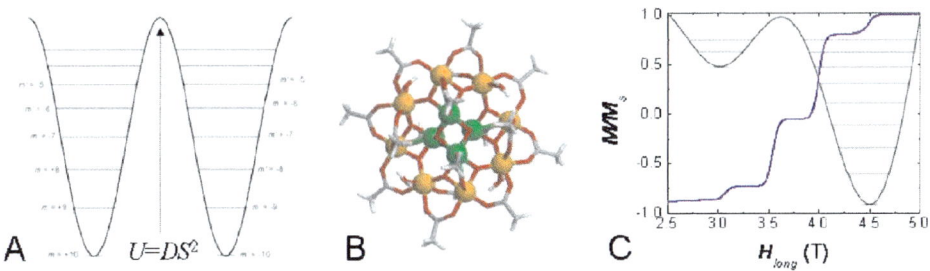

Fig. 1 (a) Energy representation of the anisotropy barrier of Mn$_{12}$-acetate, showing the energy levels corresponding to 2S+1 opposite spin projections along the easy magnetic axis of the molecule. (b) Graphic representation of single molecule magnet Mn$_{12}$. (c) Magnetization curve of Mn$_{12}$-acetate at T = 0.6 K. The jumps are due to resonant quantum tunneling at the resonance fields. The background shows the potential energy wells tilted by the action of a longitudinal field.

An effective spin Hamiltonian that captures the basic physics of SMMs is

$$\mathcal{H} = -DS_z^2 - g\mu_B \mathbf{S} \cdot \mathbf{H} + \mathcal{H}_A + \mathcal{H}_{int}. \qquad (1)$$

The first term on the right-hand side of Eqn. (1) is the uniaxial anisotropy, which results from spin-orbit interaction and generates an easy magnetic axis for the magnetic moment of the molecule. This term generates an energy barrier (DS^2 = 60 K in Mn_{12}-acetate) that separates opposite spin projections (see Figure 1a). The second term is the Zeeman energy resulting from the interaction of the spin of the molecule with an externally applied magnetic field. The other terms correspond to transverse anisotropies (H_A) and inter- or intra-molecular interactions such as dipolar, exchange, or hyperfine interactions (H_{int}). A magnetic field applied along the easy axis of the molecule tilts the potential energy wells, favoring the spin projections in the direction of the field (see the background of Figure 1c). There are certain values of the field, known as resonances ($H_k = kD/g\mu_B$, with k = 0,1,2...), for which opposite levels at both sides of the barrier coincide in energy (the background of Figure 1c shows the structure of spin levels for the resonance k = 6). At low temperatures, the magnetization hysteresis curve shows steps (accelerations of the magnetic relaxation) coinciding with these resonant fields (blue line in Figure 1c). This phenomenon was first observed in 1996 by Friedman and coworkers in a Mn_{12}-acetate SMM and interpreted in terms of resonant quantum tunneling of the magnetization (QTM) [7].

1.1.2. Antiferromagnetics Molecular Wheels

Antiferromagnetic (AFM) molecular wheels are another class of molecular magnets where an even number of transition metal ions (e.g. Fe) with finite spin organize themselves forming a closed ring in which next neighbor AFM interactions result in a S=0 ground state at low temperature. In antiferromagnetic systems the exchange interaction gives rise to pronounced spin quantum dynamics. In these systems, the Néel vector is expected to tunnel coherently for a time longer than that expected for their ferromagnetic counterparts. Several AFM molecular wheels have been synthesized and characterized to date [4-6]. In these structures, axial anisotropy aligns the spins along an axis perpendicular to the wheel plane. Due to the spin dynamics induced by antiferromagnetic exchange, the classical Neel states are not energy eigenstates. Rather, the spins are expected to tunnel jointly between the two degenerate classical ground state configurations.

AFM wheels, as well as SMMs, grow forming single crystals in which a high degree of monodispersity can be achieved. However, the zero spin value of their low temperature ground state makes it difficult to study the dynamics of the Néel vector using the accustomed characterization tools of magnetic materials. To overcome this problem, one can replace one ion of the ring by another element of different spin value, which results in a non-zero spin ground state for the molecule, hence detectable by standard magnetic characterization techniques, such as EPR spectroscopy or magnetometry. Several examples of this approach have been reported recently [10,11] and illustrate the possibilities of molecular magnets for the study of quantum dynamics under a flexible tailoring of the intrinsic properties of the quantum system.

1.2 Sources of Decoherence in Single Crystals of Molecular Magnets

There are several sources of decoherence in solid state systems, the main sources that prohibit measurement of quantum dynamics in molecular magnets are neighboring dipolar interactions and the influence of nuclear moments in transition metal ions.

Most of previous studies have been exclusively based on experiments where the quantum dynamics of the magnetic moment of the molecules is governed by incoherent quantum tunneling, meaning that the characteristic time of the experiment is longer than decoherence time of the system. Short decoherence times are mainly due to interactions between SMMs within the crystal and interactions of the electronic spin with the nuclear spin of neighboring ions within the molecule. The dispersion of structural parameters such as molecular symmetry and anisotropy complicate the study of the intrinsic quantum dynamics of the spin of an individual molecule. Also the coupling between the nanomagnets and their local crystalline environment substantially masks decoherence processes that disturb the quantum states of the system. This situation has limited the understanding of the molecular spin dynamics and the sources of decoherence. Particularly, intermolecular dipolar interactions within the crystal, which considerably shorten the coherence times in concentrated samples.

In this work I have used two methods to try to overcome these types of quantum decoherence processes: (1) synthesis of crystals of doped wheels, with the magnetic molecules spaced far apart from each other and (2) trying to study a individual magnetic molecules (not in a crystal).

CHAPTER 2: THE SINGLE CRYSTAL APPROACH

2.1 Magnetic Dilution of Single Crystals

I will discuss detailed electron paramagnetic resonance (EPR) study of single crystals of AFM Fe_{18} wheels in which, as a result of controlled doping with Gallium, only a small percentage of the molecules possesses a magnetic ground state, in effect minimizing intermolecular dipolar interactions while still preserving the monodispersity of system. The latter is evidenced by the cleanness of the EPR spectra, showing absorption peaks associated to transitions between the spin levels of the resultant magnetic $Fe_{17}Ga$ wheels that narrow as the dilution level i.e., intermolecular distance increases.

I have studied single crystals of a new family of $Fe_{18}pd_{12}pdH_{12}O_2CR_6NO_{36}NO_{36}$ ferric wheels, which are antiferromagnetic rings of 18 Fe^{3+} ions[17]. This Fe_{18} compound crystallizes as large cubic crystals in rhombohedral space group $R3$ and in high overall yields 85%. The Fe^{3+} $S=5/2$ ions were controllably substituted with Ga^{3+} $S=0$ to produce antiferromagnetic $Fe_{17}Ga$ wheels with a ground-state spin value of 5/2. The molecular structure and spin configuration of a wheel singly doped with a Ga^{3+} ion is shown in Figure 2a. The Ga^{3+} concentration was varied during the synthesis to statistically produce well dispersed $S=5/2$ wheels diluted within a sea of antiferromagnetic isostructural molecules see Figure 2b. Note that magnetic dopants have been previously used as spin markers to help determine the intrinsic magnetic anisotropy and intramolecular exchange coupling constants in single crystals of antiferromagnetic molecular wheels i.e., Abbati *et al.*[18] made use of Fe doping at a 10% concentration to characterize Ga_6

wheels, but there is no previous study on the effect of magnetic dilution on the dynamics of crystalline samples of molecular nanomagnets.

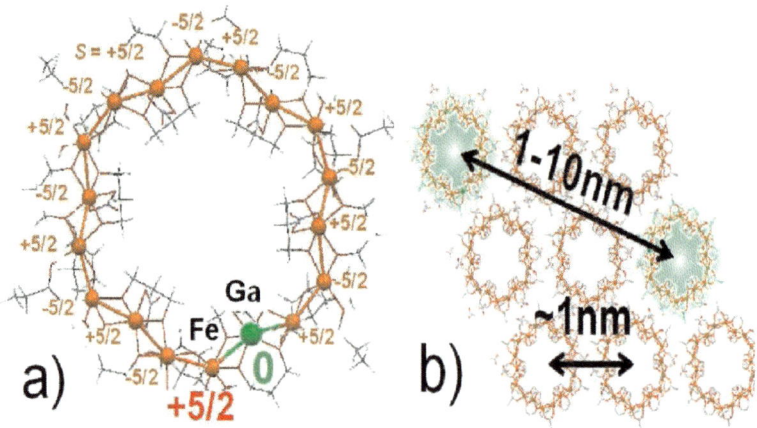

Fig. 2: (a) Molecular structure of a Fe_{18} (Fe^{3+}, S = 5/2) antiferromagnetic wheel doped with a single Ga^{3+} (S = 0) ion in one of the six most probable positions (circled in blue). This doping leads to spin-frustration and a magnetic ground state S = 5/2 at low temperature. (b) Representation of the magnetic dilution achieved in a single crystal of Fe_{18} molecular wheels by controlling the average separation between magnetic molecules through Ga-doping

In the present case, Ga:Fe doping concentrations ranging from 0.5% down to 0.005% result in average distances between magnetic molecules varying from 2 up to 8 nm. Elemental analysis carried out on the studied samples confirmed the mentioned doping ratio for concentrations over 0.1%, below which the technique is not sensitive enough. The significance of the control of the concentration lies in the ability to diminish the dephasing due to dipolar interactions between the magnetic molecules, which is inversely proportional to the cube of the magnetic intermolecular distance ($\tau_d \sim f_0 S^2/d^3$ with $f_0 \sim$ 100 MHz). In our case, dipolar dephasing times up to ~1 μs are expected for the lowest concentration (c = 0.005%), shown below in Figure 3 is the dipolar dephasing time plotted as a function of the distance between two dipole moments.

The dipolar dephasing time which is associated to the dipolar coupling between the spin of the molecule and neighboring molecules is a function of the intermolecular separation. For the typical separation of molecules in a single-crystal of SMMs, d ~1nm, the dipolar dephasing time will limit the coherence time of the system to a few nanoseconds which is too fast to gain any information about a given system. In order to overcome this problem and perform measurements of the decoherence time of SMMs up to several microseconds we have diluted samples in order to increase the distance d between neighboring molecules.

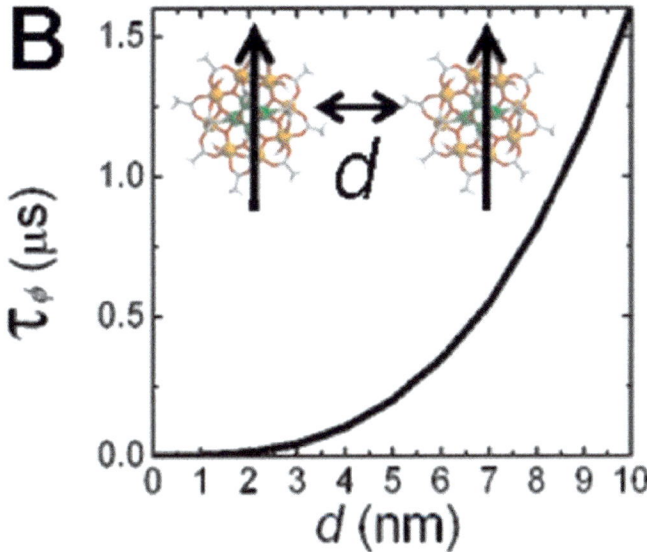

Fig. 3: Phase relaxation time originated by dipolar coupling with surrounding molecules as a function of the intermolecular separation

2.2 Development of Microstrip Resonators for High-Frequency EPR Spectroscopy Studies

In order to measure EPR of very dilute molecular wheel systems I needed an extremely sensitive device. For this I have used on-chip microstrip line resonators with a high quality factor and very strong coupling to the crystal system of study.

2.2.1 Microstrip Line Resonators

Microstrip resonator lines have been used for EPR spectroscopy for a number of years and provide an excellent method to achieve high AC fields at a crystal sample. Resonators provide a high enough Q factor that they can be used to study EPR of small crystals of molecular magnets.

I have used optical lithography to create high efficiency microstrip resonators to work in the 1-50 GHz frequency range and designed to be sensitive to SMMs single crystals down to $(50\mu m)^3$ in size. The geometry and resonant characteristics of these resonators provide an efficient conversion of the microwave power into ac magnetic field at the sample; allowing large ac field amplitudes to be applied to the sample at low temperatures. In order to achieve higher quality factor values I have also investigated the use of stencil masks as opposed to optical lithography, since the photo-resist used in optical lithography limits the thickness of metal you can deposit and by increasing the thickness of metal you can increase the sensitivity of the resonator line.

Figure 4 is a sketch of a typical microstrip resonator designed to work at frequencies in the 5-30 GHz range. The geometry of the resonator (i.e. width of the line, w, and thickness of the dielectric substrate, h, separating the line from the metallic bottom plate) has been calculated to match the impedance of the microwave coaxial lines (50Ω). As seen in the sketch, a central line of length L is isolated from the microstrip feed lines by two gaps g_c (coupling gap) and g_t (transmission gap). The coupling gap, g_c, is designed to critically couple the resonator to the feed line, resulting in quality factors on the order of order of 100. High quality factors are required to generate high microwave magnetic fields at the sample.

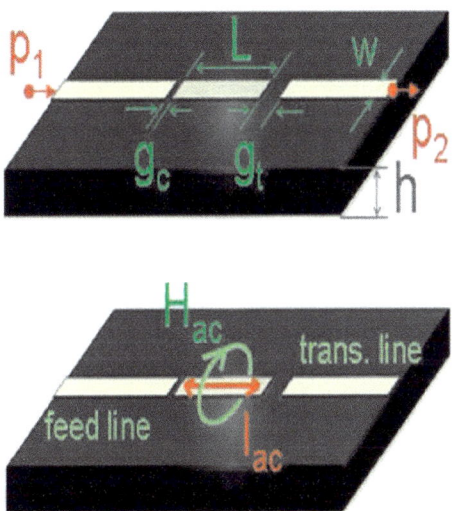

Fig. 4: Sketches of a microstrip resonator indicating the characteristic geometrical parameters (up) and functionality (down).

Shown in Figure 5 (dashed line in the left hand graphic) is the calculated reflection parameter, S_{11}, of a reflection microstrip resonator ($g_t = \infty$) with a fundamental resonance frequency of 15GHz. The superficial ac current density at resonance is shown in Figure 5 (line a), being maximum at the center of the resonator, this corresponds to the microwave magnetic field being maximum. Reflection resonators are not convenient for low temperature because the coaxial lines that run to the sample are broken at different places to thermalize the cryostat, this causes reflections and standing waves which generate oscillations of the reflected power (S_{11}) masking the response of the resonator. To solve this, I have used transmission resonators including a second feed line on the other side of the resonator separated by the transmission gap, g_t. The second line increases the losses and lowers the quality factor.

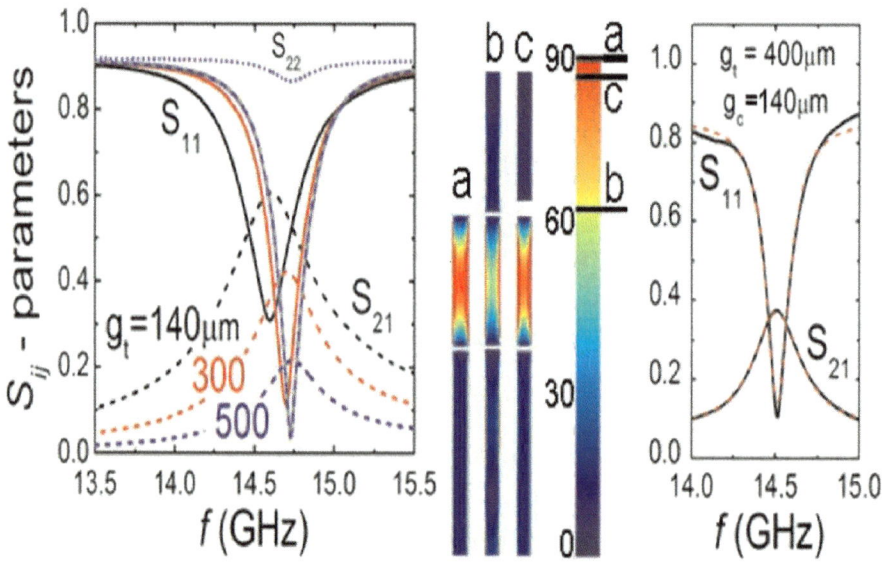

Fig. 5: Left: Calculated reflection and transmission S-parameters of a transmission microstrip resonator for four different sizes of the transmission gap, g_t = 140, 300, 500µm and ∞ (reflection resonator, undistinguishable from the 500µm resonator response). Center: Color-code plot of the ac current density (in A/m) at resonance in three different resonators; $g_t = \infty$ (a), $g_t = g_c = 140$µm (b) and $g_t = 400$µm (c). Right: Measured response of a microstrip resonator fabricated on a GaAs substrate, with the following parameters: h = 500µm, L = 3mm and w = 500µm.

The black lines in Figure 5 show the reflection S_{11} (solid) and the transmission S_{21} (dashed) parameters of a transmission resonator with a transmission gap, $g_t = g_c = 140\mu m$. Note that 30% of the power is reflected back, the peak's width is substantially larger leading to lower ac current density at resonance (line).

By increasing the transmission gap we can avoid this distortion. Figure 5 (graphic on the left) shows the S_{ij}-parameters of two other resonators with different transmission gap sizes. Both the transmission and the reflection at resonance decrease (Q increases) upon increasing the transmission gap. Over a certain gap size the reflection response of a transmission resonator is indistinguishable from that of a reflection resonator while the transmission is still large enough to measure. The measured response of a real transmission resonator built on a GaAs substrate and gold metallic lines is shown on the right of Figure 5.

2.2.2 Microstrip Cross Resonators

I have designed resonators which allow for irradiating the sample with circularly polarized microwaves. Such a polarizer is shown in Figure 6 It consists of two line resonators that form a symmetric cross. A minimum of three lines are required to operate this resonator in transmission mode. Two microwave signals, whose relative phases are externally controlled, using a phase shifter, are applied to ports 1 and 3, while the power transmitted to port 2 is monitored.

This resonator has resonances at two different frequencies (see Figure. 6). One of them occurs at the frequency of the line resonator and corresponds to the excitation shown in the cross of the Figure 6 the other corresponds to a quadrapolar excitation of the whole cross.

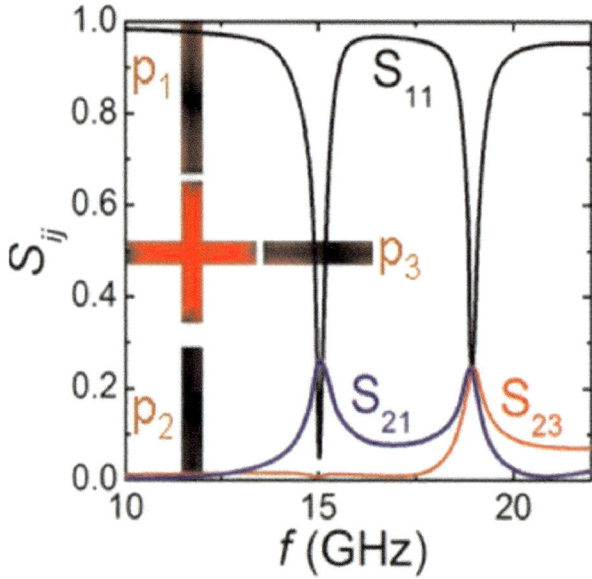

Fig. 6: Measured response of a microstrip resonator fabricated on a GaAs substrate, with the following parameters: h = 500µm, L = 3mm and w = 500µm.

In the first resonance, when excited from both ports the relative phase between the two signals controls the polarization of the microwave at the center of the cross. Varying the relative phase between the two input signals while maintaining the same power is achieved using a phase shifter and a step attenuator in the microwave lines. This polarizable microwave resonator allows the application of polarized microwave radiation at the sample at low temperature. This is not possible by polarizing the microwave outside of the cryostat because the phase is quickly lost in coaxial lines. In addition the polarizable resonator allows for selected spin transitions to be studied, since the sign of the spin transition is related to the direction of polarization (left or right) of the polarized microwaves.

2.3 EPR Studies of Diluted Fe17Ga Wheels

Using the high sensitivity microstrip line resonators fabricated via optical lithography on a GaAs wafer as described above I performed low frequency EPR experiments on diluted systems of AFM Fe wheels [10]. Figure 7 shows an optical micrograph of a resonator designed to work with its fundamental mode at 10 GHz. A typical $Fe_{17}Ga$ single crystal can be seen mounted on the 400 μm wide resonator with vacuum grease.

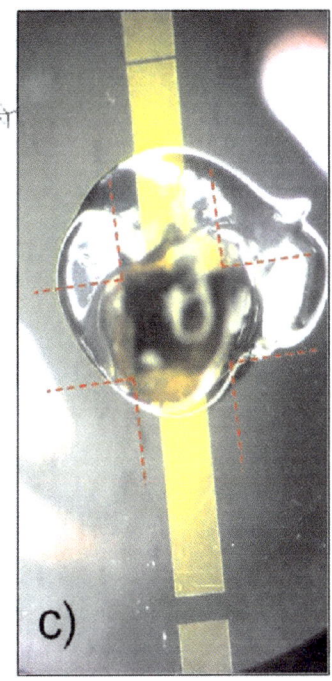

Fig. 7: Optical micrograph of a microstrip resonator, A typical Fe18:Ga crystal can be seen mounted on the 0.4-mm-wide microstrip line resonator with vacuum grease.

The EPR spectra ($f = 9.7$ GHz) of a Fe$_{18}$ single crystal doped at a 0.005% Ga:Fe concentration measured at different temperatures ($T = 0.3$, 1.8 and 4.5 K) with the magnetic field oriented along the wheel axis (axial anisotropy axis) is presented in Figure 8. The data have been averaged over 30 field sweeps due to the weak signal obtained for such a low-concentration crystal, showing four peaks in the spectra. Three peaks, with their indices denoting $|S_z|$, originate from transitions between levels with the same $|S_z|$ value, where $S_z = -S...+S$ are the projections of the spin of the molecule onto the z-axis (i.e. wheel axis). The peak labeled "(-1/2,3/2)" and "α" corresponds to transitions between states of different spin projections (from spin level $S_z = -1/2$ to $S_z = +3/2$, in this particular case).

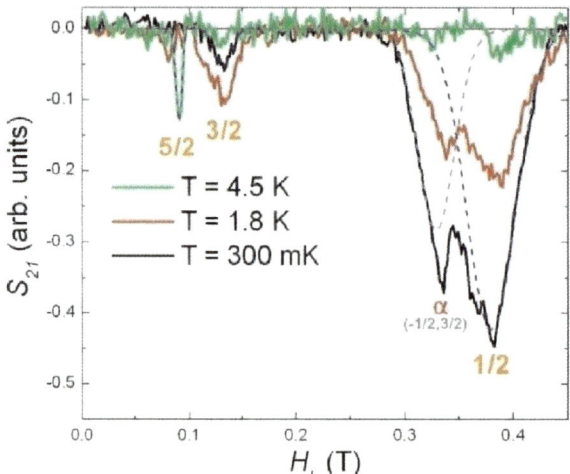

Fig. 8: 9.7 GHz EPR spectra as a function of the longitudinal magnetic field, H_L, recorded at different temperatures on a single crystal of Fe$_{18}$ molecular wheels doped with Ga at a concentration of 0.005% (Ga:Fe), which corresponds to an average separation between magnetic molecules of ~8 nm. The behavior of the EPR peaks with temperature is indicative of a positive uniaxial anisotropy parameter, $D > 0$. Three peaks, labeled 1/2, 3/2 and 5/2, correspond to transitions between levels with the same $|S_s|$. A third peak, labeled (-1/2,3/2), corresponds to transitions between states of different spin projection.

In addition, a weak contribution from a Fe mononuclear impurity, which is located at the crystal surface and is also found in pure Fe_{18} AFM crystals, is expected at this field value (H = 0.33 T) corresponding to g=2.

Figure 9 shows the fit of the spin Hamiltonian in Eqn. 2 with orange arrows showing |S| transitions seen in the longitudinal EPR field sweep.

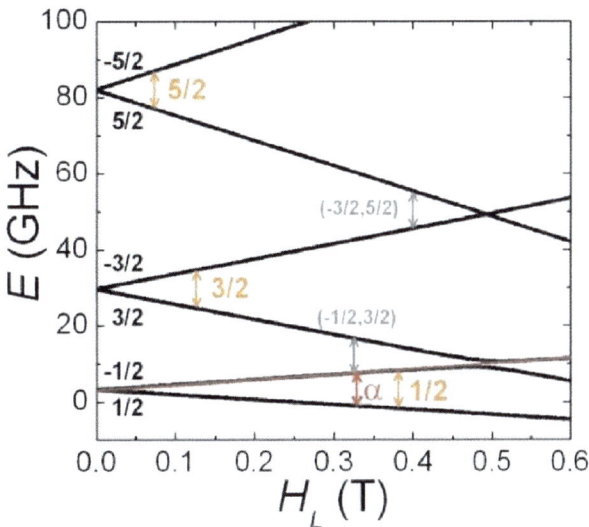

Fig. 9: Energy level diagram calculated by direct diagonalization of Eqn. 1 with magnetic field aligned with the wheel z-axis, the indicated transitions correspond to the peaks observed in Figure 8.

The behavior of the peaks with temperature indicates that the $|S_z| = 1/2$ spin projections are the molecule ground states, which determines a positive axial anisotropy parameter ($D > 0$) in the spin Hamiltonian. Lorentzian functions have been used to fit the EPR peaks (dashed blue lines in Figure.8) at the lowest temperature they become visible. The largest peak at low temperature corresponds to the −1/2 to +1/2 transition, and has a linewidth of ~400 G. The −3/2 to +3/2 and −5/2 to +5/2 transition peaks have linewidths of ~135 G and ~40 G, respectively. Observation of the -3/2 to +3/2 and -5/2 to +5/2 transitions, which should be forbidden for a molecule with six-fold symmetry, is indicative of a symmetry lowering effect that produces off diagonal terms in the spin Hamiltonian. In addition, the fact that the peak widths do not follow the 5:3:1 ratio associated to the spin levels involved in each transition can also be understood in terms of a transverse anisotropy term which affects (curves) the Zeeman energy of the levels. Moreover, the shift of the "1/2" peak from 0.33 T, expected from a linear Zeeman splitting between opposite spin 1/2 projections, to 0.38 T is also indicative of a presence of a transverse anisotropy term curving the field behavior of the spin levels, this can be attributed to the substituting Ga ion, which certainly breaks the high symmetry of the AFM wheel, most likely generating a two-fold symmetry within the plane of the molecule, causing a small second order E term in the spin Hamiltonian.

The small linewidths (>40 G) of these transitions illuminate one advantage of our single crystal dilution method. In the case of the dilute solution sample of Cr_7Mn rings with anisotropy studied by Ardavan et al [12], the overall linewidth was 5000 G due to orientational averaging in the frozen glass sample, complicating the elucidation of the transitions excited at a given field. Similarly, linewidths over 500 G were found by Bertaina et al [13], in samples of isotropic magnetic molecules, where the degree of inhomogeneity of the system is intrinsically lower. In contrast, through a determination of the spin Hamiltonian parameters, it is possible to know the spin energy landscape for a single crystal oriented in any magnetic field direction.

EPR spectra as a function of a transverse magnetic field (applied perpendicularly to the wheel axis) have been recorded for the same frequency and temperatures (see Figure 10). Two peaks are clearly observed at ~0.14 T ($\Delta_{1/2}$) and ~0.37 T ($\Delta_{3/2}$). These peaks are generated by transitions between superposition states of opposite spin projections, split by an energy Δ_{ij} ($i = -j$, with $i = 1/2$ and $3/2$, respectively) by the action of the transverse field. The weak contribution of the paramagnetic impurity is observed as a small peak (α) at ~0.33 T. The overall decrease of the magnitude of the peaks at higher temperatures, also observed in longitudinal field EPR spectra recorded up to 15 K suggests the presence of excited states. This is confirmed by high frequency EPR experiments carried out in both doped and undoped crystals of Fe_{18} wheels, where higher frequency (>50 GHz) absorption peaks are associated to excited states ($S > 0$) of the AFM molecules.

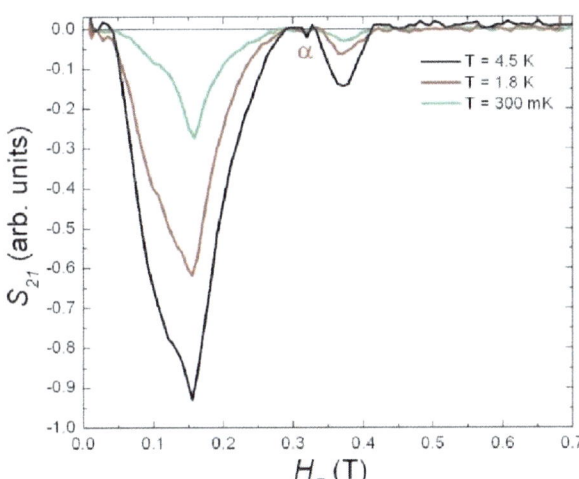

Fig. 10: The transverse field EPR spectra recorded at the same frequency and temperatures. The observed $\Delta_{1/2,-1/2}$ and $\Delta_{3/2,-3/2}$ peaks correspond to transitions between the split superposition states associated to the spin projections $|S_z| = 1/2$ and $|S_z| = 3/2$, respectively[19]. The paramagnetic impurity is observed as a weak peak (α) at ~0.33 T.

I have associated the larger widths of the transverse field EPR transitions seen in Figure 10 (in comparison to the longitudinal field spectra) to the introduction of a single Ga^{3+} ion, which, as said above, imposes a two-fold symmetry on the ring, generating a second order transverse anisotropy term in the Hamiltonian. Because we sample an ensemble of molecules in the crystal rather than one molecule at a time, based on statistics, the EPR spectra is not expected to reveal the two-fold symmetry associated to each individual molecule. In contrast, broader transverse field EPR peaks are expected to result from the fact that the relative orientation between the second order anisotropy axis and the applied transverse field will depend on the situation of the Ga ion within the wheel, and this will vary for different molecules within the crystal, which will resonate at slightly different transverse field values.

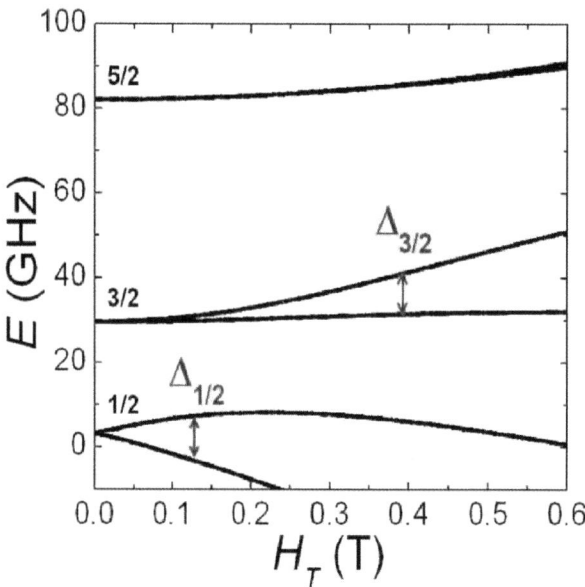

Fig. 11: Energy level diagram calculated by direct diagonalization of the Hamiltonian given in equatio.2 (see text for parameter values) with the magnetic field aligned perpendicular the wheel z axis. All indicated transitions correspond to the peaks observed in Figure 10.

The spin Hamiltonian which best describes the Fe$_{17}$Ga molecular magnet is

$$\mathcal{H} = DS_z^2 + E(S_x^2 - S_y^2) - \mu_B \mathbf{S} \cdot \hat{g} \cdot \mathbf{H}, \qquad (2)$$

where the first term is the uniaxial anisotropy, and sets the predominant magnetic behavior of the molecule depending on the sign of D. The second term represents second-order transverse anisotropy imposed by the symmetry-lowering effect of the substituting Ga ion. The last term is the Zeeman energy associated to the coupling between the molecule spin and the external field.

The observed results are better explained with the Hamiltonian of Eqn. (2) using $D = +0.63$ K, $E = 10$ mK, $g_z = 1.9$ and $g_\perp = 2.3$. Note that a positive value of D imposes an easy magnetic plane parallel to the wheel plane. The given values of the Hamiltonian parameters (particularly in the case of the transverse terms, E and $g_\perp$) need to be taken just as rough estimations (average) due to the uncertainty in the position of the Ga ions for different wheels. Nevertheless, the accuracy in the determination of these parameters is of secondary. Of special relevance is the fact that the spin-Hamiltonian parameters allow us to construct the spin energy levels diagram for any magnetic field orientation. The arrows in figure 11 highlight the resonances observed in Figure 10. The symmetry lowering effect imposed by the Fe-Ga substitution can be easily seen in Figure 11 as a curvature of the levels due to degeneracy breaking at the anti-crossing between levels of opposite spin projections (see level repulsion at ~0.48 T).

Angular dependence EPR measurements reveal a magnetic axial symmetry along the principal rotational axis of the molecular wheel, with no definite modulation of the peaks positions in the presence of transverse fields applied at different angles within the wheel plane, see Figure 12.

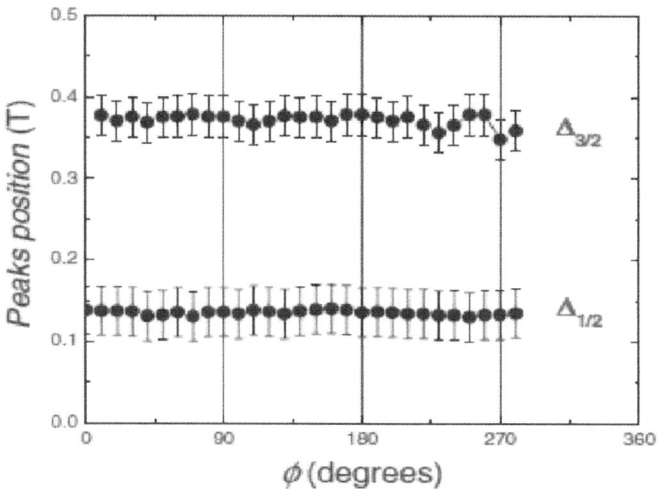

Fig. 12: Position of the 9.7 GHz EPR absorption peaks as a function of the angle of application of the transverse field within the plane of the wheel. The experiments have been performed at T=300 mK on a single crystal of Fe18 molecular wheels doped with Ga at a concentration of 0.005% Ga:Fe.

In order to understand the effect of the magnetic dilution of our crystals we have averaged the EPR spectra over 50 measurements recorded at 5 K while sweeping the longitudinal field through the peak associated to transitions between the $|S_s| = 5/2$ spin projections. Note that this is the peak whose width is least affected by the transverse anisotropy (far from the anti-crossing points). The resulting EPR absorption peaks (with normalized area) corresponding to the studied Ga:Fe-concentrations are shown in Figure 13.

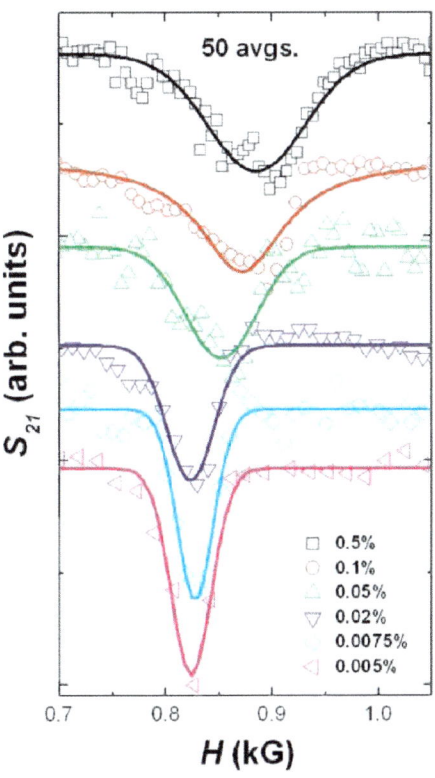

Fig. 13: Behavior of the EPR peak corresponding to the Sz = ±5/2 transition at 9.7 GHz as a function of the Ga:Fe doping concentration (0.005%-0.5%).

The peak narrows and slightly moves to low fields for decreasing concentrations until saturating below c ~ 0.01%. This can be clearly seen in Figure 14, where the width of the peak is plotted as a function of the concentration. The times next to the data in Figure 14 are estimates of the transverse relaxation times extracted from the peak width (varying from 1.2 to 3.7 ns).

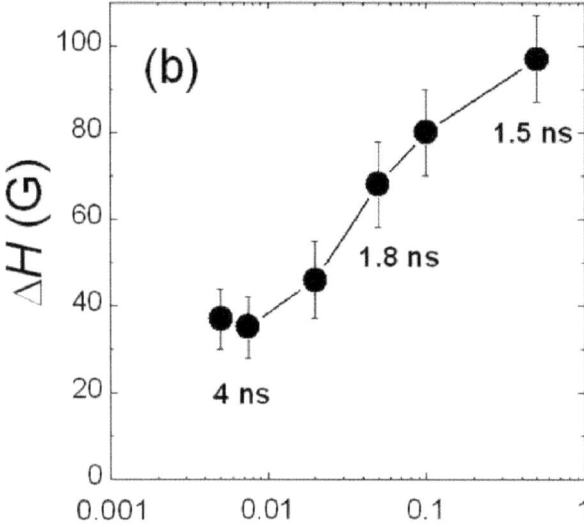

Fig. 14: Peak's width versus concentration. The peak narrows for low concentrations and saturates at ~37 Oe below 0.01% Ga:Fe concentration. The dephasing times extracted from the peak's width are shown below some of the data.

In Figure 15, the average distance between magnetic molecules within the crystal as a function of the doping concentration is shown. The times given next to the data are the dipolar dephasing times calculated from the average distance (varying from 9 ns to 1.8 μs). The difference between the estimated and calculated transverse relaxation times, together with the saturation of the peak's width and position for low concentrations, evidences that the peak broadening is of an inhomogeneous nature, mostly due to dipolar interactions between the magnetic molecules within the crystal. Note that dipolar interactions are expected to broaden the peaks, but also to change the average position of the EPR peaks when combined with other dispersion mechanisms intrinsic to the system, such as distribution of anisotropy parameters or anisotropy axes orientations. In our case, where different molecules present different orientations of the transverse anisotropy axes, the reduction of the dipolar (transverse) fields felt by the molecules at low doping concentrations weakens the curvature of the levels and, correspondingly, shifts the resonance condition to lower fields, as observed in the data.

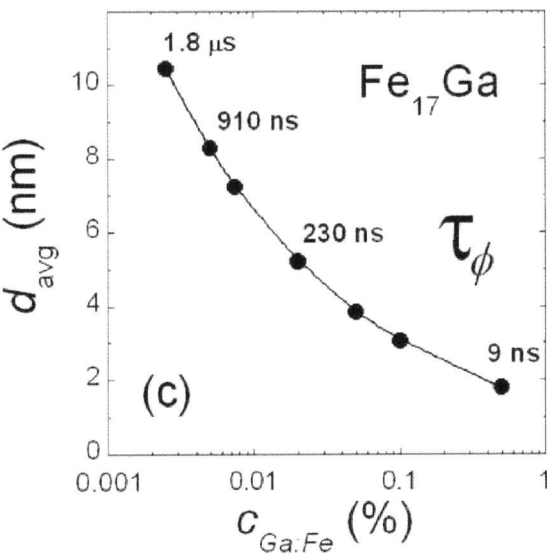

Fig. 15: Calculated average distance between magnetic molecules (Fe17Ga) as a function of the concentration. The calculated dipolar dephasing time associated to the intermolecular distance is provided for some of the data. The saturation of the peak width together with the difference between the calculated and estimated dephasing times indicate that the origin of the peaks broadening is inhomogeneous.

Our results exemplify how the magnetic dilution of single crystals decreases the degree of inhomogeneity of the system while preserving the crystalline monodispersity. This method allows the study of the quantum dynamics of molecular magnets in solid-state form without the restrictions imposed by the spin-bath polarization approach[15,16].

2.4 Observation of Quantum Mechanical Spin Selection Rules in the SMM Mn3

Symmetry enforces spin selection rules in quantum systems including QTM, only allowing tunneling transitions with a change of spin projection, $|\Delta m_S|$, equal to the order of the creation/annihilation spin operators associated with the transverse anisotropy terms in the Hamiltonian. Thus, e.g., molecules with S_4 symmetry (e.g. Mn$_{12}$tBuAc; tetragonal crystal lattice), should only tunnel when $|\Delta m_S| = 4n$, where n is an integer; and molecules with C_3 (e.g. Mn$_3$ [20-26]; trigonal lattice) or D_2 (e.g. Fe$_8$; triclinic lattice) symmetries should only tunnel when $|\Delta m_S| = 3n$ and $2n$, respectively. Surprisingly, in all experimental studies on SMMs to date, QTM is found to occur at all resonances, regardless of the spin selection rules imposed by the molecular symmetry, of which only subtle manifestations have been reported so far [27,28]. In previous manifestions of spin selection rules in SMMs all QTM resonances are visible. In fact, resonances forbidden by symmetry are found to be comparable in magnitude to the allowed ones. Evidence is extracted mainly from subtle changes in magnitude between them [28], or from their distinct behavior upon application of a transverse magnetic field [27]. There are many possible reasons for this. One is that intrinsic disorder (coming from solvent loss, ligand disorder or crystalline defects) lowers the local molecular symmetry so that an applied longitudinal field induces a distribution of transverse magnetic field components [19-31]. Other possibilities include the dynamic fields from nuclear spins, and dipolar interactions [32]. What has been lacking so far is a clear-cut experimental demonstration of the role of molecular symmetry in QTM, in which all of these extraneous effects are either eliminated or accounted for.

The crystalline quality of this particular molecule is demonstrated by the unsurpassed sharpness of the X-ray diffraction and EPR absorption peaks [20,21]. We observe a sequence of very sharp QTM steps, where one resonance appears only at high temperature. We have shown that the high temperature relaxation at this resonance is associated with tunnel transitions involving excited states with $|\Delta m_S| = 3n$, as dictated by the C_3 symmetry of the molecule, which forbids tunneling from the lowest metastable state at low temperature. In addition, a rotation of the local zero-field-splitting (ZFS) tensors following the tilts of the Jahn-Teller axes of the Mn^{III} ions, in combination with intermolecular dipolar interactions, accounts for the observed behavior in all QTM resonances, including transitions forbidden by the molecular symmetry.

We have studied two complexes **[NEt$_4$]$_3$[Mn$_3$Zn$_2$(salox)$_3$O(N$_3$)$_6$Cl$_2$]** and [NEt$_4$]$_3$[**Mn$_3$**Zn$_2$(salox)$_3$O(N$_3$)$_6$**Br$_2$**], henceforth Mn$_3$-Cl and Mn$_3$-Br, respectively [20,21]. The metallic cores of these complexes are comprised of a μ_3-oxo-centered triangle of Mn^{3+} ions and two capping Zn^{2+} ions located above and below the Mn$_3$ plane, resulting in a rigid trigonal bipyramidal structure. The diamagnetic Zn^{2+} ions and bulky [NEt$_4$]$^+$ cations isolate the Mn$_3$ magnetic core from intermolecular magnetic interactions, as evident from the absence of significant intermolecular contacts and the 10.30 Å minimum separation between Mn^{III} ions in neighboring molecules. Both complexes crystallize in the trigonal space group $R3c$ as racemic mixtures of C_3-symmetric chiral molecules (with equal population of molecules with opposite chirality, rotated by 27 degrees about the C_3 axis with respect to each other). Neither structure contains solvate molecules, which is quite rare for SMMs and likely explains the extremely high resolution spectroscopic data (solvents evaporate easily, causing disorder [33]). Ferromagnetic exchange interactions between Mn^{3+} ions are propagated by the central μ_3-oxo ion and through the coordinating oxime, resulting in a molecular spin $S = 6$ ground state at low temperature. These structural and crystallographic properties differentiate Mn$_3$-Cl and Mn$_3$-Br from other ferromagnetic Mn$_3$ triangles, each of which possesses appreciable intermolecular interactions, low molecular symmetry, or co-crystallized solvate molecules [34-38]. We observe only subtle differences in the magnetic behavior of Mn_3-Cl and Mn_3-Br; we will thus consider them identical in the context of the discussion below.

Magnetization hysteresis measurements were carried out on sub-millimeter size single crystals of both Mn_3-Cl and Mn_3-Br, using a high-sensitivity micro-Hall effect magnetometer [39] in the temperature range of 0.3-2.6 K. Figure 16 shows the first derivative of the magnetization (for Mn_3-Cl) plotted versus the longitudinal magnetic field, H_L (along the easy anisotropy axes of the molecules), at different temperatures. Narrow peaks corresponding to the $k = 0, 1, 2$ and 3 QTM resonances are clearly observed at almost regular field intervals ($\Delta H \sim 0.85$ T). Resonance $k = 1$ (0.85 T) is not visible at the lowest temperatures in Figure 15; it appears only raising the temperature above 1.5 K, when it appears suddenly at a lower field value (0.80 T). To the best of our knowledge, this is the first occasion in which a QTM resonance (e.g. $k = 1$) within a series of resonances is found to be absent, while lower and higher resonances are observed. As we show below, this constitutes definite evidence for the influence of spin selection rules for QTM in a SMM.

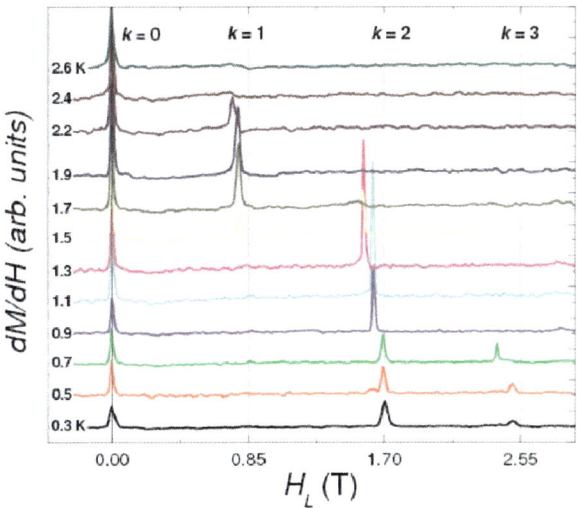

Fig. 16: Field derivative of the magnetization curves obtained for a Mn3-Cl single crystal at different temperatures, with the field swept at 1 T/min along the easy axes of the molecules.

The observed shift of all resonances to lower fields with increasing temperature indicates a transition from the pure quantum tunneling regime, in which the relaxation occurs from the ground spin state, to thermally activated tunneling between excited states [40]. The fact that the resonances associated with excited states appear at lower field values is indicative of a fourth order uniaxial anisotropy term in the Mn$_3$ Hamiltonian (i.e. $B_4^0 \hat{O}_4^0$), coming from a relatively weak exchange interaction constant (J) between the manganese ions in the single-ion spin Hamiltonian given by [42]:

$$H = \sum_i \left(\vec{s}_i \cdot \ddot{R}_i^T \cdot \ddot{D} \cdot \ddot{R}_i \cdot \vec{s}_i - \mu_B \vec{s}_i \cdot \ddot{g} \cdot \vec{B} \right) + \frac{1}{2} \sum_{i,j (i \neq j)} \vec{s}_i \cdot \ddot{J} \cdot \vec{s}_j . \qquad (3)$$

Here the first term represents the magnetic anisotropy of the *i*-th ion, $\ddot{D}$ being the ZFS (diagonal) tensor given by $D_{xx} = e$, $D_{yy} = -e$ and $D_{zz} = -d$, with *d* and *e* the uniaxial and second order transverse anisotropy parameters, respectively. $\vec{R}_i$ is the Euler matrix specifying the anisotropy axes of the three Mn ions in the molecule, defined by the Euler rotation angles θ_i, ϕ_i and φ_i. The second term is the Zeeman coupling to the applied magnetic field, and the last term is the exchange interaction between neighboring ions. The positions of the QTM resonances observed in both quantum and thermally activated regimes (Figures 15-18) can accurately be accounted for using the following set of parameters: $s_i = 2$, $d = 4.2$ K, $e \sim 0.9$ K, isotropic $g = 2$ and $J = -4.88$ K. The single-ion second order anisotropy (e) is needed to explain the observed QTM rates. Similar d and e values have been reported for Mn^{III} ions in the literature (see, for example, Ref. [42]). We additionally rotated the anisotropy axes for the three Mn^{III} ions such that $\theta = 8.5°$ (with $\varphi_i = 0$), and $\phi_1 = 0$, $\phi_2 = 120°$ and $\phi_2 = 240°$, in order to account for the local tilts of the Jahn-Teller axes and to preserve the C_3 symmetry [20,21].

Figure 17 shows the magnetization curve obtained at 300 mK for the Mn_3-Cl single crystal shown in Figure 16. From the measured changes in magnetization at the resonances, we can obtain a rough estimate of the tunnel splittings associated with the superposition of different spin states at each resonance.

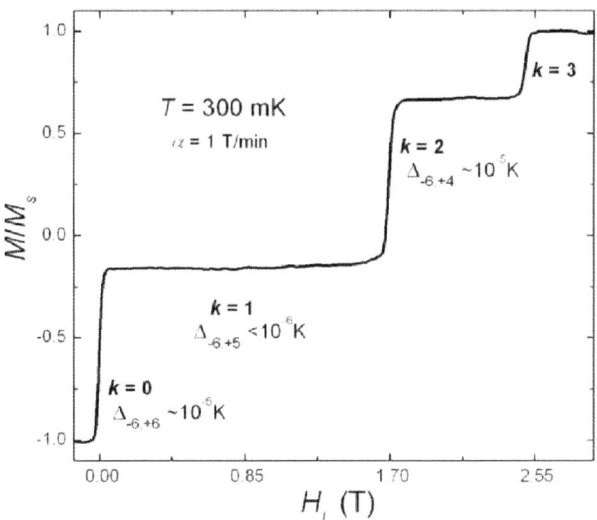

Fig. 17: Magnetization versus longitudinal magnetic field recorded at 0.3 K in a Mn3-Cl single crystal. The indicated tunnel splitting values are order of magnitude estimates from the change of magnetization at each resonance.

At resonance $k = 1$, the splitting ($\Delta_{-6,+5} < 1\times10^{-6}$ K) is found to be one order of magnitude smaller than those at resonances $k = 0$ ($\Delta_{-6,+6} \sim 7\times10^{-6}$ K), $k = 2$ ($\Delta_{-6,+4} \sim 1\times10^{-5}$ K) and $k = 3$ ($\Delta_{-6,+3} > 1\times10^{-5}$ K). Note that the values given for the tunneling splittings are estimates using the Landau-Zener formalism, and should be considered as lower limits for the low field sweep rates used in the experiment, for which reshuffling of dipolar fields during relaxation can have a significant effect.

Interestingly, the absent resonance ($k = 1$) can clearly be observed by decreasing the field sweep rate or by applying a transverse field. Figure 18a shows the field sweep rate dependence of resonance k=1 while Figure 18b shows the growth of the $k = 1$ peak as a function of the transverse field magnitude.

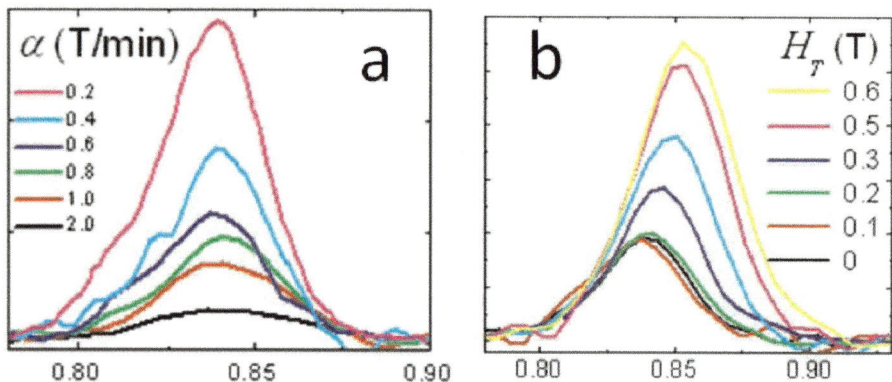

Fig. 18: a) The sweep rate dependence of resonance k=1, b) The dM/dH peak at resonance k = 1 upon application of a transverse magnetic field.

Figure 19 shows the QTM probability as a function of the transverse field. The curvature of the probability at $k = 1$ near zero magnetic field is indicative of a saturation caused by intermolecular dipolar interactions, whose magnitude (~250 G) can be estimated from the width of the peaks.

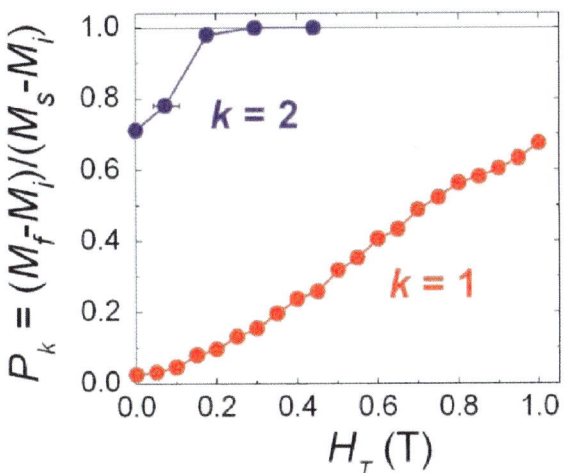

Fig. 19: The QTM probability increases monotonically at resonances k = 1 and 2 with increasing transverse field.

The C_3 symmetry of the Mn$_3$ complexes should only allow tunneling between states on opposite sides of the barrier with spin-projection (m_S) values differing by a multiple of three. According to this selection rule, the only resonances observable below the crossover temperature separating the pure quantum tunneling regime from the thermally activated one should be $k = 0$ (tunneling from $m_S = -6$ to $m_S = +6$, $\Delta m_S = 12 = 4\times3$) and $k = 3$ (tunneling from $m_S = -6$ to $m_S = +3$, $\Delta m_S = 9 = 3\times3$). Figure 20 shows the energy of the m_S levels as a function of the longitudinal field, calculated from exact diagonalization of the Hamiltonian in Eqn. 3. The first allowed QTM transition for each resonance is indicated at the anti-crossings between the respective spin levels (blue squares). Note that the lowest level-crossings involving the $k = 1$ and $k = 2$ resonances are degenerate (red circles in Figure 20), since they involve transitions that are not a multiple of three (i.e. $\Delta m_S = 11$ for ground state at $k = 1$ and $\Delta m_S = 10$ and 8 for the ground and first excited states at $k = 2$). This explains why resonance $k = 1$ only becomes visible at high temperatures in Figure 15, since an excited state needs to be populated for the QTM to take place (i.e. the transition from $m_S = -5$ to $m_S = +4$, $\Delta m_S = 9 = 3\times3$). Consequently, the temperature dependence of resonance $k = 1$ constitutes firm evidence for the spin selection rules imposed on QTM in a SMM.

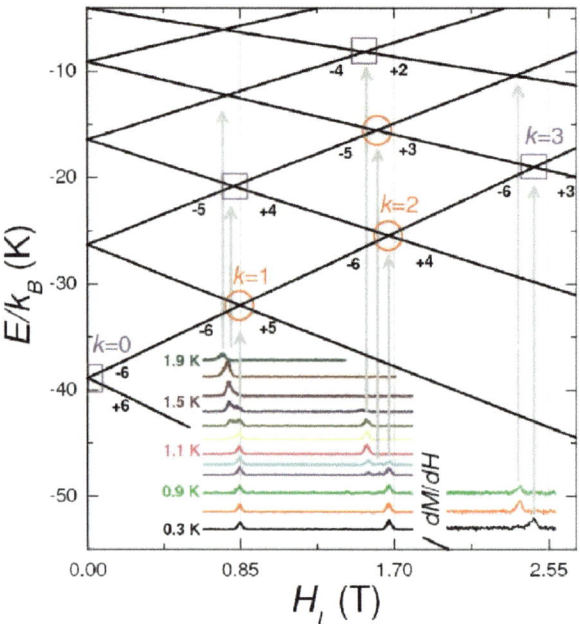

Fig. 20: Energies of the Mn3 mS levels as a function of the longitudinal magnetic field. Symbols highlight level crossings between different spin states, indicating the degeneracies (red circles) and avoided crossings (blue squares) expected for the C3 symmetry of the structures. Inset: field derivatives of the magnetization curves (dM/dHL) measured for a Mn3-Br single crystal at different temperatures, with field sweep rates of 0.05 T/min (data below 1.7 T) and 0.2 T/min (data above 1.7 T).

Following the same arguments, and contrary to the experimental observations, one should expect resonance $k = 2$ to be absent at low temperatures as well (see Figure16). This inconsistency can be understood in terms of tilting of the Jahn-Teller axes of the manganese ions within the molecule. This can clearly be observed in Figure 20, which shows the dependence of the ground state tunnel splittings on the magnitude of the transverse field for all resonances observed in the experiment (k=0-3), calculated via exact diagonalization of the Hamiltonian in Eqn.(3) using the parameters given above for two different situations: a) With the Jahn-Teller axes aligned with the z-axis, i.e. $\theta = 0$ (thin lines in Figure 21); and, b) including a tilting of the Jahn-Teller axes of each ion out of the crystallographic z-axis (molecular easy-axis), given by the Euler angle $\theta = 8.5°$ (thick lines in Figure 21), determined from X-ray crystallography data [20,21]. The observations in Figure 21 are essentially insensitive to the orientation of the transverse field within the hard anisotropy (x-y) plane over the range of parameter space explored in this investigation. This agrees with the absence of any modulation in both the dM/dH peak magnitudes and EPR absorption peak positions (to within the resolution range of the techniques)

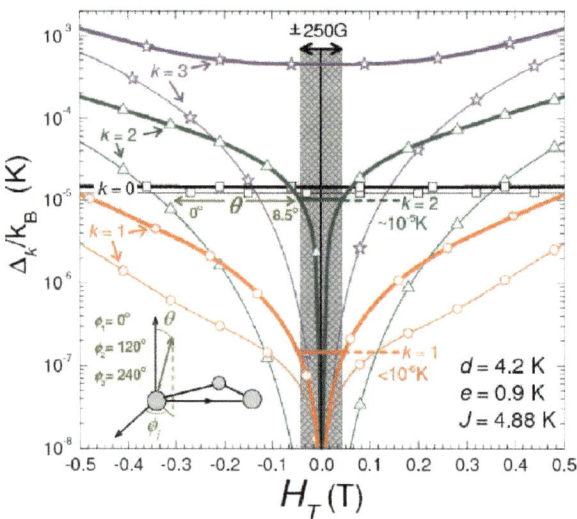

Fig. 21: Ground state tunnel splittings associated with resonances k = 0 (black squares), k = 1 (red circles), k = 2 (green triangles) and k = 3 (blue stars) as a function of the transverse field, HT, with the Jahn-Teller axes aligned along the z-axis (thin lines) and tilted θ = 8.5o away from the z-axis (thick lines). A few symbols per curve have been added to help in their identification. The grey shaded region in the vicinity of zero transverse field represents the strength of the dipolar magnetic field (~250 G) felt by the Mn3 molecules within the single crystal. The horizontal lines within the grey region indicate the order of magnitude of the splitting values attained at resonances k = 1 and k = 2 for a transverse field of 250 G.

The tilting is represented by a sketch in Figure 21. In both cases, resonances $k = 1$ and $k = 2$ are degenerate in the absence of a transverse field ($\Delta_{k=1} = \Delta_{k=2} = 0$). If the Jahn-Teller axes are not tilted, large transverse fields ($H_T > 0.2$ T) are required to bring the tunnel splittings up to the magnitudes observed in the experiment. However, the inclusion of a tilting of the Jahn-Teller axes by 8.5° has a profound influence on the transverse field behavior of the ground-state splittings at resonances $k = 1$ and $k = 2$. This effect is particularly significant in the case of resonance $k = 2$ (observed at the lowest temperature), for which a splitting magnitude on the order of 10^{-5} K is achieved for fields below ~250 G, while the ground-state splitting at resonance

$k = 1$ remains more than an order of magnitude smaller for the same range of transverse field values. As shown above, intermolecular dipolar interactions can provide magnetic fields (~250 G) which are strong enough to induce a tunnel splitting in the $k = 2$ resonance of the level observed in the experiment. Meanwhile, their effect on the $k = 1$ resonance is nearly two orders of magnitude weaker, thereby explaining the absence of this QTM step in our studies of carefully aligned crystals (see Figure 17).

These results exemplify the remarkable influence of the molecular symmetry on the magnetic relaxation of molecular nanomagnets and provide the first pristine demonstration of spin selection rules on the QTM for a SMM. The observed behavior must be attributed to the extremely high crystalline quality of these complexes, enabling deep insights into fundamental quantum behavior that were previously impossible. Of special significance is the remarkable finding that a rotation of the ZFS tensors of the individual ions (in a manner consistent with the crystallographic symmetry) has a profound effect on the transverse field behavior of the tunnel splitting in resonances forbidden by the molecular symmetry. We have shown that a small Jahn-Teller axis tilt (8.5°) is sufficient to increase the tunnel splitting value for the $k = 2$ resonance up to an observable level for transverse field magnitudes commonly provided by intermolecular dipolar interactions or as a result of weak disorder. Note that the ZFS tensors are almost never parallel in *real* structures and, according to our results, this combined with dipolar fields and/or disorder may be the ultimate reason behind the apparent absence of spin selection rules in previous studies of QTM in SMMs.

CHAPTER 3: THE SINGLE MOLECULE APPROACH

3.1. Physics of SET Devices

The charge state is intertwined with the internal electronic states of magnetic molecule, and SET devices provide a method to study the relation between electron dynamics and magnetic levels in a SMM. As an example, a paramagnetic Co-terpyridinal complex has been incorporated into a SET and its transport properties studied under external magnetic fields [43]. It was shown that the external field and the molecular spin levels directly affect the conductance of the SET. This experiment not just provided some important new results, but also introduced a new "transport spectroscopy" technique for probing magnetic excitations. SMMs will provide new insights into the interplay between conduction electrons and magnetic states of individual molecules. Transport through a SMM will be influenced by the molecule's spin states and will influence QTM and quantum coherence properties.

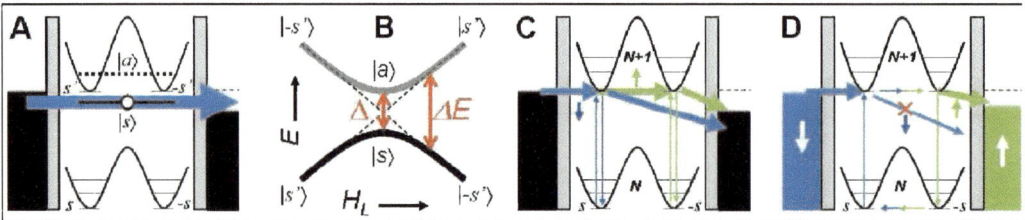

Fig. 22: The content of these diagrams is discussed in the text.

QTM is governed by the tunneling splitting (of magnitude Δ) separating symmetric $|S\rangle$ and antisymmetric $|A\rangle$ superpositions of states of an SMM with opposite spin projections, while single-electron conduction through a SET is determined by the coupling of the molecule to the leads, the magnitude of the coupling is given by $\Gamma = \Gamma_s + \Gamma_d$, where Γ_s and Γ_d, are tunneling rates between the molecule and the source and drain leads, respectively. The effect of QTM on the transport through an SMM-based SET will be dependent on the relative magnitudes of Δ and Γ, as well as on the intrinsic decoherence time of the system T_2. In the weak coupling, weak decoherence regime, when $\Delta \gg \Gamma$ and $\Delta \gg 1/T_2$, a large tunnel splitting will allow the states $|S\rangle$ and $|A\rangle$ to directly contribute to the conduction through the device (see Figure 22a). Consequently, these states will be distinguishable in the nonlinear response of the device through their dependence on an external magnetic field. Figure 22b shows the effect of a longitudinal magnetic field on the spin levels of the SMM at a QTM anticrossing resonance point. Opposite spin projections are degenerate at resonance ($H_k = k$ D /gμB) but a transverse field breaks the degeneracy and produces a tunnel splitting Δ which then curves the linear field dependence of the Zeeman energy at the vicinity of the resonance according to $\Delta E^2 = \Delta^2 + (2g\mu_B M H_z)^2$, where M is the eigenvalue of S_z. Excitation lines in the SET differential conductance related to these magnetic levels will show such curvature when as a function of longitudinal magnetic field. The opposite regime, when either coupling or decoherence are strong, is defined by $\Delta \ll \Gamma$ or $\Delta \ll 1/T_2$. With $\Delta \ll \Gamma$, the electrons time inside the molecule is shorter than the time necessary for the magnetization to tunnel across the anisotropy barrier of the SMM. When $\Delta \ll 1/T_2$, e.g., hyperfine interactions will destroy the coherence of the spin tunneling. Therefore, the electron will not see the tunnel splittings of the SMM and all electronic transitions will occur between states on the same side of the anisotropy barrier (see Figure 22c). However, there are certain ways to elude this limitation and observe the effect of QTM in this regime. The first one is a theoretical proposal by Leuenberger and Mucciolo [44] that is based on the modulation of the Kondo effect by the Berry-phase oscillations of the tunnel splitting associated to destructive

QTM interference. For the Kondo effect to occur the coupling between the molecule and the leads must be strong and the tunnel splitting must be large in comparison to the thermal energy of the system. The second one is a theoretical proposal by Gonzalez and Leuenberger [45] to use ferromagnetic leads to study the effect of QTM (and the Berry phase) on polarized current in the sequential tunneling regime. Typical values of Γ in molecular SETs fluctuate from MHz up to a few GHz. Magnetic fields can be employed to vary the tunnel splitting from H_z to hundreds of GHz, allowing tuning of the conduction regime.

3.2. Development of SET Devices

3.2.1 Single-Electron Transistors

Electronic transport properties of individual molecules have received considerable attention over the last several years due to the introduction of single-electron transistor (SET) devices [22-34]. A scheme of a three-terminal SET is represented in Figure 23a. The molecule is placed between three electrode leads (source, drain, and gate). The electrostatic coupling between the molecule and the leads is of a capacitive nature, since the ligands surrounding the molecule act as insulating barriers. The capacitances depend primarily on the molecule/lead distances, but also on the molecule's ligand composition. While placed far from the gate, ligands attach the molecule to source and drain electrodes and provide tunneling barriers for the electron to move in and out of the molecule. Therefore, an electric current can flow between the source and drain electrodes through a sequential tunneling process.

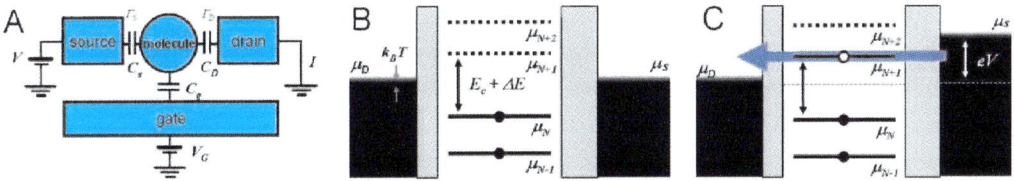

Fig. 23: (A) Schematic representation of a three terminal SET. (B) Energy representation of the electrostatic levels of the molecule with respect to the Fermi energies of the source and drain electrodes. N represents the molecule on its neutral charge state, being N+1 and N-1 its first reduction and oxidation states, respectively. E_c is the charging energy of the molecule. (C) The effect of a bias voltage on the energy configuration of the molecular SET. Note that the first exited state (N+1) of the molecule is now available for conduction, since it lies within the Fermi energies of the electrodes.

Figures 23b and 23c represent the energy landscape of an ideal molecular SET. The black regions on the sides represent the electron Fermi seas in the source and drain electrodes, with μ_S and μ_D being the Fermi levels of the leads. The grey blocks represent the tunnel barriers between the molecule and the source/drain leads (note that these barriers can be quite asymmetric since the disposition of the molecule with respect to the electrodes may vary). The charge states of the molecule are represented by the horizontal lines in between the barriers. The highest of all occupied states (solid lines) represents the molecule with N electrons and an electrochemical potential μ_N. The first (empty) excited state is separated by an energy $E_c + \Delta E$, where ΔE is the molecular electronic level spacing and E_c is the energy necessary to add one electron into the molecule (charging energy or redox potential). Conduction through a molecular SET only occurs when a molecular electronic level lies between the Fermi energies of the leads. A bias voltage V applied between the source and the drain moves the Fermi level of one of the leads by $|eV|$. For small bias voltages, $|eV| < E_c + \Delta E$, no current flows though the device because the excited molecular levels are not available to accept conduction electrons (Figure 23b). This regime is known as Coulomb blockade. As the bias voltage is further increased, excited states open new conduction channels through the device. Abrupt and discrete changes in the current through the SET will be obtained every time a new molecular level becomes energetically accessible. The voltage values at which these current steps occur can be tuned by a potential applied to the gate electrode, V_G, which moves the molecular states with respect to the Fermi levels of the electrodes.

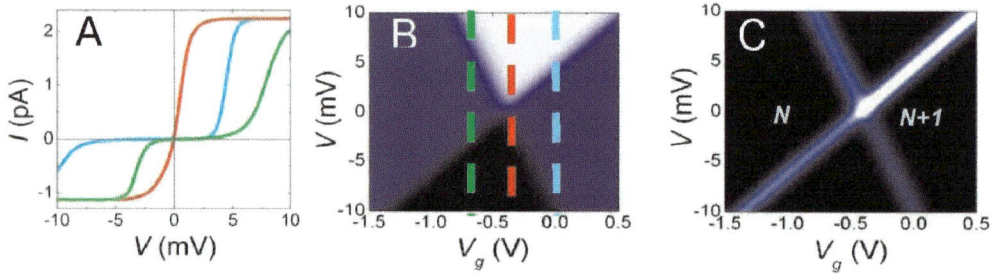

Fig. 24: (a) Calculated current for an ideal single-state molecular SET as a function of the bias voltage, V_{BIAS}, for different gate voltages, Vg. b) I and c) dI/dV contour-plots of a molecular SET calculated assuming asymmetric barriers and tunnel rates. The dashed color lines in b correspond to the I-V curves shown in a).

The suppression of current near zero bias (source drain voltage) is known as the Coulomb blockade, this behavior and the characteristic conduction behavior of an SET are illustrated in Figure 24, where calculations of the current flowing through an ideal SET are presented. Figure 24a shows the typical *I-V* curves observed in a SET device at three different gate voltages. Discrete steps are observed whenever a new excited state is available for conduction. Contour plots of the current (Figure 24b) and the differential conductance (Figure 24c) as functions of the bias and gate voltages show the characteristic diamond structure representative of Coulomb blockade. Figure 24c illustrates two consecutive charge states of the molecule (N and $N+1$) that are separated by excitation lines. These lines intersect at the point where the gate voltage equals the charging energy of the molecule. The dI/dV plot of an SET also reveals the energetic level structure of the molecule and is a very powerful spectroscopy technique to study the energy landscape of individual molecules. As in other spectroscopic techniques, the position, the shape, the magnitude, and the slope (among others) of the conduction excitations display characteristics that are both intrinsic (i.e., the electronic nature of the molecule) and extrinsic (i.e., the disposition of the molecule with respect to the electrodes) of the system under study.

3.2.2. Fabrication of SET Devices

Measuring electrical conduction through an individual molecule is not a straightforward task. The main difficulty is the impossibility of obtaining electrodes separated by just a few nanometers (< 3 nm) using conventional lithographic techniques. There are several approaches that try to circumvent this problem. Scanning probe microscopy techniques, such as STM or AFM have been widely used to study the conduction through individual molecules deposited on a metallic surface. However, this approach requires sophisticated instrumentation, extensive expertise in low-temperature techniques, and cannot make use of a gate electrode to study the different charge states of the molecule. Alternative approaches involve the on-chip fabrication of nm-size gap electrodes. There are two such methods: The *mechanical break junction technique* [46] (where a suspended nanowire is broken by stress applied to the substrate), the *electromigration technique* [47] (where a metal nanowire is broken by a current).

As shown in the following, I have been able to manufacture operative three-terminal SETs with a 1-3 nm gap separation between source and drain electrodes, and a 2-3 nm separation to the gate electrode. The devices are fabricated in a multi-step process that involves optical lithography, electron beam lithography (EBL), high-vacuum metal deposition (Au, Al, as well as other metals), and electromigration breaking. The process follows the detailed description provided in the Ph.D. thesis of Jiwoong Park at UC Berkeley [47].

Figure 25 contains a selection of images of our SET devices. Figure 25a shows a 3-D view of the chipdesign containing five sets of six nanowires. A photograph of the entire chip is shown in Figure 25b. A total of 30 nanowires is obtained on each chip. In every run we fabricate 16 chips, which makes a total of 480 nanowires per run. (Having a very large number of nanowires is of crucial importance due to the low yield of the technique employed to generate molecular SETs.) The device fabrication procedure is the following (see Figure 25a). A thin (20 nm) Au layer patterned according to the lead geometry is deposited on a Si/SiO_2 wafer (1 μm thick SiO_2 top layer to prevent leaks). A thick (200 nm) Au layer (yellow) is then deposited on top to decrease the overall resistance of the leads and help in wire-bonding the contacts to the chip holder. Note that the ends of the leads (orange) are left thin. A narrow Al gate electrode (grey), 20 nm thick, is deposited while maintaining the sample cold (77 K) during evaporation. This helps to obtain a smooth flat Al surface. The Al is left at atmospheric conditions overnight to obtain a thin (2-3 nm) Al_2O_3 layer. The gate is connected to two electrode leads (upper-left and lower-right regions in Figure 25a). All steps up to this point are done exclusively with optical lithography.

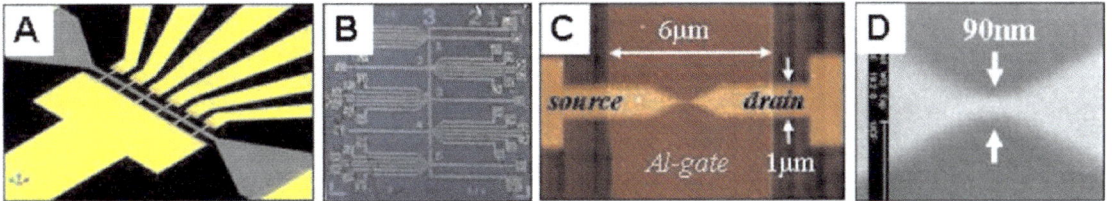

Fig. 25: A) Design perspective and B) photograph of our SET device. The chip on B is 1´1 cm2 and contains 35 SETs grouped in five sets of six transistors each, as shown in A. The substrate is pure silicon with a one micrometer silicon oxide layer on top. C) AFM image of a zoomed area in the chip where the wire nanoconstriction and the Al gate are shown. D) SEM image of the gold nanowire.

The final step in the process is the patterning of thin (18 nm) gold nanowires by means of state-of-the-art EBL. Figure 25c shows an AFM image of one of the nanowires deposited on top of the Al-gate. A zoom on the nanowire center (Figure 25d) obtained by SEM shows a nanowire approximately 200 nm long and 90 nm wide. All components of the circuit and steps in the process were tested repeatedly until an optimal design was obtained.

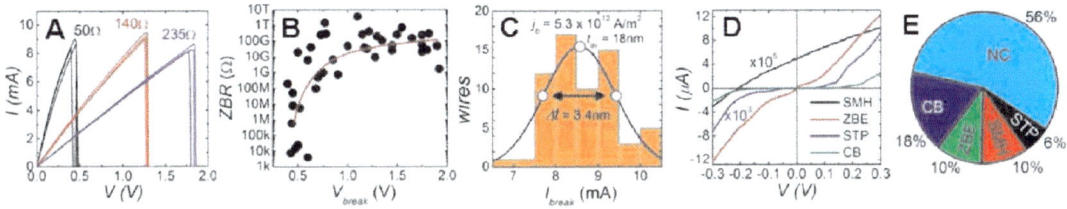

Fig. 26: a) IV curves recorded during electromigration-induced wire breaking for three circuit resistances. b) Zero bias resistance (ZBR) as a function of breaking voltage. c) Distribution of breaking currents for a circuit resistance of 235 W. d) Different IV curves observed after wire breaking: Asymmetric smooth curves (SMH), curves with zero bias enhancement of the current (ZBE), curves with steps (STP), and curves with current suppression at low voltages consistent with Coulomb blockade (CB). e) Pie chart showing the statistics of the IV curve types found after the breaking of over 60 nanowires. NC corresponds to curves where no current was observed after breaking.

Figure 26a shows the electromigration-induced breaking ($T = 4$ K) of several nanowires. The current across the nanowires increases linearly with the voltage until the rupture point, when a gap is created and, consequently, the current abruptly drops to zero at a given *breaking voltage*, V_{break}. A resistor (20, 110, or 205Ω) in series with the nanowire (30 Ω) is used to tune the breaking voltage. Figure 26b shows the breaking voltage dependence of the tunnel resistance of the gap measured at low bias voltages (the *zero-bias resistance*, ZBR) of several wires after breaking. The increase of the resistance with the breaking voltage is associated to the gap formation during the electromigration-induced breaking of the nanowire, which is mainly determined by the current density. It is well known that for a given current density the higher the voltage the wider the gap. The tunnel resistance varies between 10 kΩ and 100 GΩ for most wires. According to previous studies [47,48], this huge variation corresponds to just about a 1 nm variation in gap size. This means that our gap sizes are in the 1-3 nm size range. The current density necessary to break a Au nanowire is estimated to be $j_b = 5\times10^{12}$ A/m^2 [50] and the breaking current I_{break} necessary to achieve this characteristic current density depends on the cross section area of the nanowire. Therefore, the breaking voltage is determined by the total resistance of the circuit, namely $V_{break} = R_T I_{break}$, where $R_T = R_{nw}+R_{series}$. Consequently, a series resistor can be used to control the breaking voltage, and therefore engineer the size of the gap. The histogram in Figure 26c shows the distribution of breaking currents centered at around 8.5 mA. Considering a cross section area of 90 nm (width) × 18 nm (thickness) for our nanowires, we obtain $j_b = 5.3\times10^{12}$ A/m^2, which is in excellent agreement with the value given in Ref. [50]. If we associate the change in breaking currents ($\Delta I_{break} \sim 2$ mA) to variations of the nanowire thickness, this indicates a 3.4 nm thickness variation along the four-inch long Si wafer used to fabricate the chips. This variation is likely due to the dispersion of the Au evaporation beam along the wafer (solid angle). In Figure 26d we show four different types of *I-V* curves found after the formation of the gap and Figure 26e shows the statistics of each of the curves together with the wires showing *no current* (NC) after breaking (most likely due to an extra-large gap formation). The *I-V* curves of broken wires can then be grouped according to the following classification: a) *CB*, curves with current suppression for low bias voltages consistent with the Coulomb blockade effect; b) *STP*, curves showing abrupt changes of current (steps) consistent with quantization of the conductance; c) *ZBE*, zero bias enhancement of the conductance consistent with low-temperature Kondo effect; and d) *SMH*, smooth asymmetric curves not crossing $I=V=0$. Both the performance and statistics that we have obtained in our first SET devices are similar to those of other, more experienced groups [47,48].

Figure 27 shows an SEM micrograph of an electromigrated gold wire:

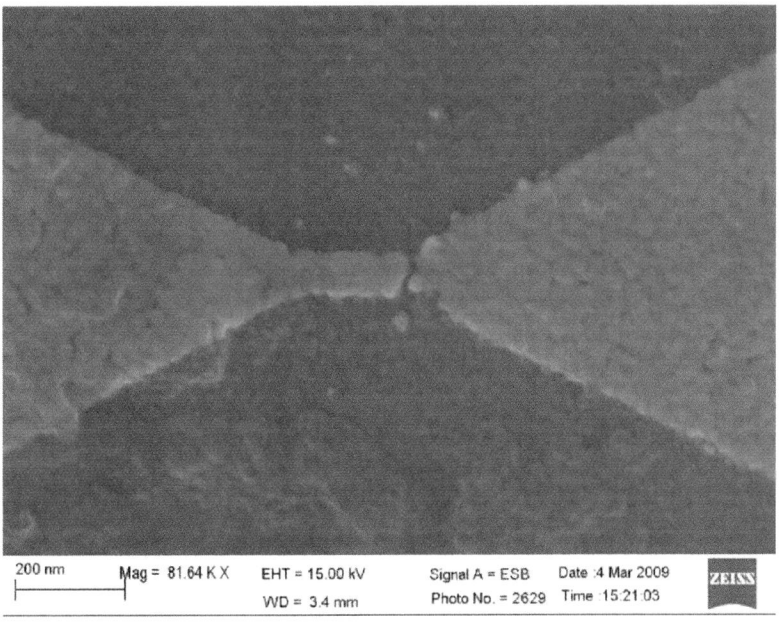

Fig. 27: Electromigrated gold wire showing a small gap and gold particles that were removed.

3.2.3. Functiontionalization of SMMs

Molecules have been prepared and functionalized in microcrystalline powders for us by the Christou group at the University of Florida Department Of Chemistry. For the beginning of this study we used Mn_{12}-acetate ($Mn_{12}O_{12}(O_2CMe)_{16}(H_2O)_4$) provided to us in microcrystalline form. This family of Mn_{12} is S = 10 and was chosen because the Christou group can replace all of the acetate groups of this compound with any other carboxylate, the Eqn. for this process is Eqn. 3 below. [49-52]

$[Mn_{12}O_{12}(O_2CMe)_{16}(H_2O)_4]$+16 RCO_2H → $[Mn_{12}O_{12}(O_2CR)_{16}(H_2O)_4]$+16 $MeCO_2H$ (3)

Using the R group in Eqn. 3 one can put whatever carboxlate group on the outside of the magnetic core of the molecule, so you can tailor your ligands to be whatever you want. We started with R groups containing sulfur since that is known to stick strongly to gold surfaces which is what are SET devices are made out of.

We have also been working more recently with a Mn_4 family of cubes, they are S = 9/2 [53-55]. The core of these molecules is [Mn_4O_3X], where X is variable and can be X= F, Cl, Br, N_3, NCO, OH, OMe, for instance.

3.2.4. Deposition of SMMs

The microcrystalline powder is then dissolved into a solution using methylene chloride as a solvent. A chip is then plasma cleaned for 30 minutes using AMPAC's plasma cleaner and placed in the solution for varying amounts of time. The molarity of the solution has to be carefully controlled using a microbalance, this provides a variable amount of molecules in solution.

In order to be even more clear about our surface coverage we have performed AFM on epitaxially grown gold surfaces in order to carefully map out the molarity of the solution and the time of deposition needed to have a nice clear understanding of how many molecules per unit area we will have on the SET devices to try to maximize the chances of having a single molecule sitting in the electromigrated nano-gap of the SET device.

Figure 28 shows $Mn_{12}O_{12}(TBP)_{12}(O_2CMe)_4]H_2O$ molecules deposited using a 0.1 mM solution for varying time, Figure 28a shows bare gold surface without molecules, 28b shows 5 minute deposition time where you can see molecules on the surface but also patches of bare gold, 28c shows 10 minute deposition time where the molecules appear to be almost an even coating of the surface and Figure 28d shows 60 minute deposition time with layered coverage of the surface.

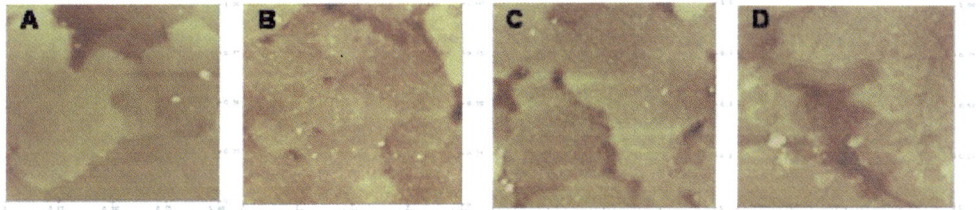

Fig. 28: a) bare epitaxial gold surface b) 5 minute deposition of $Mn_{12}O_{12}(TBP)_{12}(O_2CMe)_4]H_2O$ time c) 10 minute deposition time d) 60 minute deposition time.

3.2.5. Preliminary Results

The experimental data shown in Figure 29 were taken at $T = 4$ K in an SET covered with Mn_{12}-3tpc molecules. Figure 29a shows the stepwise I-V-curves measured at different gate voltages. Peaks in the current derivatives (Figure 29b) clearly reveal the discrete nature of the conductance through our SMM-based SET. The conduction excitations (current steps) can be easily followed in Figure 29c, where the differential conductance (dI/dV) is contour-plotted as a function of bias and gate voltages (white lines are to guide the eyes along the linear conductance excitations). The presence of multiple parallel excitations forming the characteristic diamond shape of a molecular SET indicates the crossing between two charge states (N and $N+1$) of the molecule at $V_g \sim 0.3V$ and reveals the complex nature of the excited states of this particular molecule. The excitations are found in an energy range (0-40 meV) similar to what had been observed by other groups for Mn_{12}-based SETs [56,57]. Similar behavior has been observed for other molecules in the same set of chips.

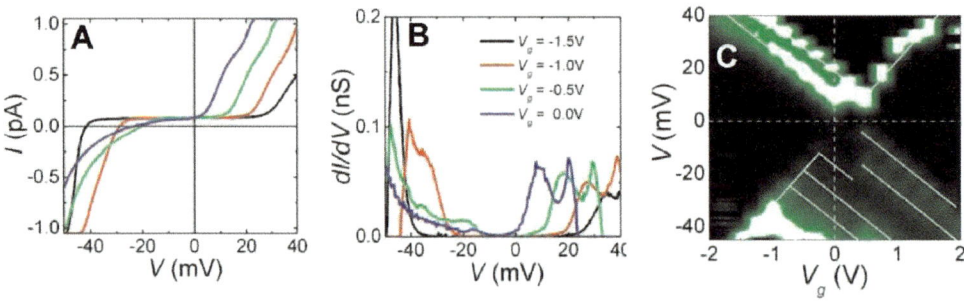

Fig. 29: a) IV curves recorded at 4K with a Mn_{12}-3tpc based SET. b) Differential conductance vs. bias voltage for the same gate voltages of B. c) dI/dV contour plot as a function of the bias and gate voltages showing the typical response of a molecular SET.

I have demonstrated the ability to fabricate functional molecular SET devices, although thus far we have not yet obtained any clear evidence of SMM behavior.

CHAPTER 4: CONCLUSIONS

Molecular magnets are an exciting solid state system showing extremely interesting physical quantum phenomena. During this thesis work I have studied primarily three methods to try to extract novel information about these systems. Using two approaches (single crystal studies and individual molecule studies) I have learned a lot of new and exciting information and have been lucky enough to have several works published in physics journals.

In the single crystal measurements I've performed the dilute crystal approach to trying to measure differences in minute amounts of chemical synthesis changes and seen how it affects EPR absorption lines by using home-designed on chip microwave resonators. These resonators allowed me to see changes in absorption peaks as a function of amount of Ga added to Fe during the synthesis of $Fe_{17}Ga$ doped Fe_{18} wheels. I was able to measure many crystals of different concentration and noticed that the crystalline environment became more inhomogeneous as more Ga was added and caused greater disorder. I think this method of trying to measure pulse EPR will work to eliminate dipolar interactions but maybe a simple system would be better to eliminate other dispersion mechanisms.

While studying the magnetization of a new simple trigonal symmetric Mn_3 crystal of single molecules we happened to find an interesting phenomenon: A missing QTM resonance that had been hinted in previous experiments. This missing resonance led to a lot of discussion in the field of SMMs and eventually led to a very interesting result. Due to the lack of solvent disorder we were able to see this missing resonance that was predicted by theory. The symmetry of the Mn_3 triangle imposes a quantum spin selection rule that says that only resonances that are multiples of three should appear in the magnetization measurements. The present results exemplify the remarkable influence of the molecular symmetry on the magnetic relaxation of molecular nanomagnets and provide the first pristine demonstration of spin selection rules on the QTM for a SMM. The observed behavior must be attributed to the extremely high crystalline quality of these complexes, enabling deep insights into fundamental quantum behavior that were previously impossible. Of special significance is the remarkable finding that a rotation of the ZFS tensors of the individual ions (in a manner consistent with the crystallographic symmetry) has a profound effect on the transverse field behavior of the tunnel splitting in resonances forbidden by the molecular symmetry. We have shown that a small Jahn-Teller axis tilt (8.5°) is sufficient to increase the tunnel splitting value for the $k = 2$ resonance up to an observable level for transverse field magnitudes commonly provided by intermolecular dipolar interactions or as a result of weak disorder. Note that the ZFS tensors are almost never parallel in *real* structures and, according to our results, this combined with dipolar fields and/or disorder may be the ultimate reason behind the apparent absence of spin selection rules in previous studies of QTM in SMMs.

We have outlined the technical details of our single electron transistor devices and demonstrated that they can be used to measure single-electron transport through individual Mn_{12} SMMs. Our current objective is to probe different SMMs at low temperatures and in the presence of high magnetic fields generated by a vector superconducting magnet which allows arbitrary orientation of the magnetic field with respect to the geometry of the SMM-based SET.

APPENDIX: CLEANROOM RECIPES

Please take notice that all of these recipes have been developed using facilities in the del Barco lab using certain chemicals and equipment with certain strength of radiation, any variance to a different UV lamp, different e-beam lithography machine or different hot plate may require drastic changes in these procedures.

Double Layer Optical Lithography Recipe

This is the most commonly used procedure for optical lithography in del barco lab, with this recipe one can make: microstrip resonators to study EPR, Hall sensors to study magnetization, and single-electron transistors.

I.   Turning on the power to the instruments

1. Flip the switch to the power strip located below the microscope on the black desk. Each of the red lights should come on. If not, turn them on individually.
2. Turn on the small diaphragm pump, which is located outside the clean room on the floor near the vacuum chamber and the gas cylinders.
3. Turn on the ultrasound bath.
    a. Press "*select options*" to go to "*set sonics.*"
    b. Use the "*set display*" button to adjust the time to 5 minutes.
4. On the black control box above the power strip flip the power on. This turns on the mask aligner. Wait about 5 minutes then press and hold the "start" button for 2 seconds to turn on the UV lamp.

II.   Pre-cleaning and preparation

1. Put on powder free latex gloves.

2. Take the mask from the box labeled "masks" in the first drawer of the desk.

3. Clean the substrate.
    a. Fill 3 clean plastic beakers ½ full with acetone, isopropanol, and ethanol respectively.
    b. Place the substrate into the acetone beaker and place the acetone beaker into the sonic bath. Hit the "on" button to begin the ultrasound cleaning for 5 minutes. The time remaining will be displayed.

c. Without allowing the acetone to dry, put the substrate into the isopropanol and sonicate again.
   d. Without allowing the isopropanol to dry, put the substrate into the ethanol and sonicate for a last time
   e. While you wait on the substrate begin cleaning the mask

4. *Clean the mask.*
   a. Turn on the N_2 gas that is against the wall outside the clean room to 40 psi. Press the trigger to the gun to check the flow.
   b. Take a clean cotton ball out of the drawer in one of the small plastic boxes to the right of the sink.
   c. Saturate it with acetone.
   d. Saturate it again with d.i. water.
   e. Clean the side of the mask on which the pattern is written with the cotton ball.
   f. Spray with d.i. water to remove loose dirt.
   g. Blow dry with N_2.
   h. Clean the opposing side the same way and blow dry.
   i. Examine the mask under the microscope at 20 X magnification. If need be, repeat the cleaning process.

5. *Install the mask into the mask aligner.*
   a. Remove the mask clamp (top square part of the mask aligner). This piece holds the mask in place during use.
   b. Place the mask, chrome side down, over the square hole above the sample holder (the chuck). Be careful not the scratch the mask.
   c. Carefully replace the mask clamp over the mask and flip the mask clamp button to the up (on) position. Check to see that the mask clamp does not move. If it does, you do not have sufficient vacuum. So, make sure the "OAI vac" valve located on the spinner side of the metal table is open (parallel).

6. Remove the substrate from the ethanol and blow it dry with N_2. Cover it with a beaker.

7. *Prepare the photoresists.*
   a. Clean the small brown vials with acetone, isopropanol, ethanol, and d.i. water then blow dry.
   b. Fill one vial ¼ full with LOR 3A and the other ¼ full with S 1813.
   c. Cover their tops with Al foil.
   d. Clean two glass pipets the same way and blow dry. Install rubber bulbs onto the pipettes and insert them into the photoresist vials by poking a hole in the Al foil.

8. *Prepare the developer.*
   a. Clean two glass beaker that is large enough to contain the substrate if you were holding one corner with tweezers the same way and blow dry.
   b. Fill one of the beakers with d.i. water and cover with Al foil.
   c. Fill the other beaker with the CD-26 developer and cover with Al foil.

9. Heat the yellow hot plate to 170 C.

10. Turn the N_2 pressure up to 60 psi on the regulator.

III. *Spin coating the substrate: double layer process*

1. Press "*program select*" on the spinner to choose program B, which will be indicated on the top left corner.
2. Turn on the N_2 flow to the spinner by opening the valve nearest the wall at the end of the metal table. If "CDA" flashes on the spinner, the gas pressure is too low.
3. Close the OAI vac valve on the left side of the metal table and open the spinner vac valve on the right side of the metal table.
4. Using tweezers, place the substrate centered over the rubber o-ring on the spinner. Press the "*vacuum*" button. The vacuum should read > 15.
5. Make sure program B is still active and cover the substrate with LOR 3A.
6. Press "*run/stop*" to begin spinning.
7. Bake on the hotplate for **exactly** 5 minutes. Cover the substrate with a beaker.
8. Remove the substrate and reduce the hotplate temp to 120 C.
9. Switch to spinner program A, cover the substrate with S 1813, and spin.
10. Turn off the N_2 gas on the regulator.
11. Bake for **exactly** 2 minutes at 120 C.
12. Set the hotplate to 130 C.

IV. Alignment and Exposure

1. Close the spinner vac valve and open the OAI vac valve. Make sure the mask clamp is well secured.

2. Lift the mask frame and place the substrate on the mask aligner chuck (the part with holes) under the portion of the mask you want to use.

3. Place the small sponge pieces at the corners of the chuck to help avoid breaking or scratching the mask.

4. Cover all of the open holes with clean glass slides and flip on the sample vacuum switch. If all of the holes are not covered, this will cause problems later. If you can hear a hissing sound, then some of the holes are open.

5. Close the mask frame and flip the mask frame switch to the up position. Now, the mask frame should be held secure by vacuum.

6. Focus the microsope on the mask portion that is just over the sample.

7. Turn the large black dial counterclockwise to raise the substrate toward the mask. If you feel some tension in the dial, then the chuck is pressing on the mask. Move the chuck back down slightly and then work your way up. Repeat this up-down process when tension is felt to avoid breaking the mask and to level the chuck.

8. Once the substrate approaches the mask, you should see a shadow appear in the microscope. Slowly raise the chuck until the shadow disappears. Now the sample is in contact with the mask and is ready to be exposed.

9. The exposure time will already be set to optimum performance. On the left of the mask aligner you will see the timer setting. Right now it is at 7 seconds.

10. Gently slide the mask aligner under the UV lamp region and either look away or close your eyes. Remember that UV light can damage your eyes.

11. After the second click, slide the mask aligner to its original position.

12. Lower the substrate using the black dial (turn clockwise) until it stops. Turn off the mask clamp and sample vacuum switches and remove the substrate.

V. Developing the pattern.

1. Dip the sample into CD-26 for about 30 seconds. Lightly move it periodically and you should see the photoresist on the exposed portion dissolve.

2. Dip it in d.i. water to stop developing.

3. Blow dry with N_2.

4. Examine the substrate by moving it around in the light. Use the microscope if you want to. If you see any rainbows in the UV exposed portion, then some of the photoresist remains. In this case develop for a few more seconds. If a lot of the photoresist remains, then you overbaked the sample and need to start over.

VI. Etching an undercut to make liftoff easier

1. Bake the dry substrate for 5 minutes at 130 C.

2. Dip it in CD-26 for one minute without moving.

3. Quickly dip it in d.i. water. Blow dry.

VII. Examine the sample under the microscope. Your pattern should exactly match that of the mask with sharp corners and straight lines.

VIII. Clean up your mess and turn off the instruments.

IX. Perform your evaporation (see other procedures).

X. Perform the liftoff.

1. Place the sample in remover PG. Over time, the excess photoresist and the metal layer on top will be removed and only the pattern will remain.

Double Layer E-Beam Lithography Recipe
This is the process in order to make < 1 um features using creol's ebeam lithography machine.

1) Prepare the sample using optical lithography or finding graphene sheets.

2) Spin coat MMA photoresist using program D.

3) Bake sample for 15 minutes at 175 C.

4) Spin coat PMMA using program E.

5) Bake sample for 15 minutes at 175 C.

6) Take it to creol, align and expose using 600 uC/cm^2, this gives 70 – 90 nm wide features.

7) Develop sample for 60 seconds in 1:3 MIBK : DI water. Stop bath in DI water then rinse with DI water.

8) Check the features using the microscope for alignment, size etc.

9) Deposit metal (don't use e-beam evaporator).

Fabrication of SETs with Al Gate Procedure

I. Chemicals

a) Rinse pipettes, beakers etc with acetone, IPA, and ethanol.
b) Prepare vials of resist.
c) Prepare CD-26 beaker and DI water beaker.

II. Wafer preparation.

a) Cut 3 inch Si-SiO_2 into a 1.8 inch square using the manual wafer cutter.
b) Sonicate the wafer for 5 minutes in Acetone, IPA, Ethanol then blow dry with Nitrogen gas.
c) Spin on Lor 3a resist using program B then bake on hotplate for 5 minutes at 175 degrees.
d) Spin on Shipley S1813 resist using program A then back on hotplate for 2 minutes at 120 degrees.

III. Alignment

a) Place wafer on mask aligner using either glass slides or cut away plastic to ensure sample vacuum. Turn on sample vacuum
b) With eye align wafer piece to the sample pattern on the mask.
b) Using contact vacuum raise wafer also you can use the dial.
c) Use micrometers to align layers.
d) Check that the power reading on the lamp is about 250.
e) Slide chuck all the way to the left and look away from the lamp.

IV. Development

a) Place wafer in beaker of CD-26 for 45 seconds, quickly place in DI water.
b) Spray wafer with DI water then blow dry.
c) Check patterning with microscope, if its good etch the undercut at 130 degrees for 5 minutes, the one minute in CD-26 followed by DI and Nitrogen blow dry.

V. Evaporate

a) Carefully place sample on sample holder and mount in chamber.
b) Layer 1: 3nm Cr followed by 25 nm Au
c) Layer 2: 3 nm Cr followed by 200 nm Au
d) Layer 3: 24 nm of Al (cool with nitrogen once at low pressure then wait 6-8 hrs to remove from vacuum).

VI. Lift off

a) Place sample in Remover PG at 80 degrees then check periodically
b) Spray with PG to help the lift-off (not on the Al layer since no Cr)

VII. Gates

a) Once lifted off check with meter or probe station whether the gates are conducting properly.
b) Allow gates to oxidize in atmospheric conditions for 12 hrs

VIII. E-beam Lithography

a) Spin on MMA resist with program D.
b) Bake for 15 minutes at 175 degrees.
c) Spin on PMMA resist with program E.
d) Bake for 15 minutes at 175 degrees.
e) Take it to CREOL

IX. Nano-wire Developing

a) Develop sample in 3 IPA : 1 MIBK
b) Rinse with IPA and blow dry
c) Check patterning under microscope if necessary develop a bit more.

X. Nano-wire Evaporation

a) CAREFULLY mount sample in chamber.
b) Evaporate 24 nm of Au.
c) Remove from chamber

XI. Nano-wire Lift-off

a) Place sample in RT Acetone and leave it over night. It helps to prop the sample up again the beaker wall.
b) Don't rush or the wires will be gone.
c) Lift sample out of Acetone and see if the gold comes off.
d) If it doesn't right away wait a little bit longer.
e) If it still doesn't come off try spraying gently with Acetone.
f) Blow dry with a very light stream of nitrogen gas.

XII. Does it work?

a) Check that the gates are still conducting
b) Check that there are indeed wires there (~70 ohm at RT) by measuring with probe station.
c) Check if there is a leak between the sources (drain) and gates with probe station.
d) If there is a leak check ALL chips to check, if all of them leak them you just wasted lots of time and money.

XIII. Measuring

a) Using the closed cycle system check a chip that has lots of wires at low T.
b) Break a wire or two using the straight voltage sweep breaking software. Note the voltage it broke at.
b) Using the feedback controlled software break wires using a critical voltage noted in the previous step.
c) If there is some conduction there immediately check for leaks to the gate.
d) Assuming the gate is working try taking some I-V curves with varying gate (-1, +1 volts say). Be mindful of the gate oxide it could crack if you up the voltage too much leaving you with a leak and the entire device being bad.

Hall Sensor Fabrication Recipe
Materials needed:

Photoresists: Shipley S1813 and LOR-3A
Developer: CD26
Remover: PG
Other Chemicals: Acetone, Isopropanol, ethanol, H_2O_2, H_2SO_4, DI-water, Hidroxid peroxide, Ammonium Hydroxide.
Metals for evaporation: Platinum (or chromium), gold and Au/Ge (88/12%).
Wafer: 2DEG GaAs/AlGaAs
Tools: glass slides, pipettes, diamond knife, lint-free paper, hot plate, ultrasound bath, spinner, mask aligner and evaporation chamber.

A. SAMPLE PREPARATION

1. Take a glass slide and clean it with ultrasounds:
   Acetone (5 min.) – Isopropanol (5 min.) – Ethanol (5 min.)
2. Blow dry with N_2 gas.
3. Take the 2-DEG wafer and cut the piece you want to use.
   a) place the wafer over a microscope glass slide with the part you want to cut off out of the slide. Cover the 2-DEG surface with lint free paper and place a second glass slide over it exactly at the same position as the first glass slide. Only that part of the wafer you want to cut off must be out of the glass slides.
   b) mark a line on the wafer with the diamond scribe.
   c) press gently at the edge of the wafer closest to you with a wood stick until breaking the wafer.
4. Take the wafer piece and clean it with ultrasounds: Use a plastic baker or a piece of Teflon if using a glass baker to avoid rupture of the wafer by the ultrasounds.
   Acetone (5 min.) – Isopropanol (5 min.) – Ethanol (5 min.)
5. Prepare two small bakers with the photoresists S1813 and LOR-3A. It is useful to cover the bakers with aluminum foil, to avoid exposure of the photoresists.
6. Clean two glass pipettes with acetone –isoproponol and DI water and blow dry.
7. Introduce the pipettes into the bakers with the photoresists by making a hole through the aluminum foil. This way, neither dust or light will ruin the photoresists during the time of the process.
8. Put a small drop of S1813 in the glass slide with the help of a glass pipette. Spread it out over an area proportional to the wafer piece area. Use only a bit of liquid.
9. Place the wafer piece over the photoresist drop on the slide and move the slide (with the wafer on) gently until observing the piece move freely over the slide.
10. **Bake it during 15 minutes at 115°C (120°C in our plate)** to harden the photoresist. Cover the sample with a glass baker during baking.

Before next step, be sure to have everything ready in the Mask aligner for exposition of the sample and the developer CD26 and DI water ready to be use in respective vials.

B. SPINNING THE PHOTORESIST-SINGLE LAYER PROCESS (prior 2DEG etching)

11. Place the sample on the spinner. Hold it with vacuum. Set the spinner to program A (2 seconds at 500 rpm followed by 30 seconds at 5000 rpm). The sample should be out of the center of rotation. Note that our spinner needs pressure (N2 gas 40psi) and vacuum (small mechanical pump) to work. Have these ready.
12. Using the previously cleaned pipette, cover the sample completely with S1813 with care in not touching the surface.
13. **Spin it for 2 seconds at 500 rpm followed by 30 seconds at 5000 rpm**. This would give you a 1.2 um layer thickness.
14. **Bake the sample for 2 minutes at 115°C (120°C in our plate)**.

C. EXPOSURE – MASK ALIGNER

15. Turn on the vacuum pump.
16. Turn on the system (Mask aligner lamp, vacuum and pressure systems).
17. Wait few minutes (~5 min)
18. Press and hold the lamp-source START (white) button until reading the status of the lamp. It should be around 250-270 W Power (constant) and 60 Volts current, and channel 2 Mode CP (constant power).
19. Place the mask in the holder (aligner) and hold it with vacuum. Before closing the mask holder, *make sure the sample holder is at its lower position*, to avoid the mask to touch it. *(These previous steps should be done well in advance)*
20. Place the sample in the sample holder and hold it with vacuum. (For our mask aligner: If the glass slide containing the sample does not cover all the holes, then other slides should be used. In this case, other (fake)wafer pieces are needed to equilibrate the mask with the sample)
21. Align the mask with the sample at the desired position. Use the microscope for this.
22. Move the sample up until making close contact with the mask. Do not force the mask, you must stop when you feel that you need to apply more force to move it.
23. Select the desired **exposition time (7 seconds)**, this should be determined ad-hoc.
24. Move the aligner with the sample under the lamp.
25. Turn on the lamp to exposure the sample.

D. DEVELOPING S1813-SINGLE LAYER PROCESS

1. Prepare pure developer CD26 in a baker and DI-water in another one. (*This needs to be done before exposure*)
2. Just after the UV exposure, take the sample with plastic tweezers and **deep it in the CD26 developer for 45 seconds**. You should see the rainbow resulting from diffraction to disappear once the photoresist layer is gone.
3. Immediately after that, **deep the sample in DI water** to completely stop developing.
4. Rinse the sample with DI water and then blow dry with nitrogen.

The sample is ready for etching.

E. ETCHING THE 2DEG

1. Prepare one vial with DI water
2. Prepare another vial with **DI-water:H_2O_2:H_2SO_4 – 160:8:1** (For this put first the water, then the peroxide and the sulfuric at the end. Use a metric vial for measuring the proportion).
3. **Deep the sample in the acid mixture during 13 seconds** (etch-depth ~110nm).
4. Immediately after that, deep the sample into DI water to stop the etching.
5. Blow dry with nitrogen.

F. REMOVING PHOTORESIST – LIFT-OFF

1. Deep the sample in acetone during a few minutes (1-2 min.)
2. Deep it in isopropanol for a minute.
3. Deep it in ethanol for a minute.
4. Blow dry.

G. OHMNIC CONTACTS – DOUBLE-LAYER PROCESS (UNDERCUT)

To pattern the contacts follow steps B, C and D for single photoresist layer process of follow instructions for double photoresist layer if you need an undercut. Either method will work well for Hall sensors. The undercut method allows a better lift-off after metal deposition. We describe here the double layer process:

1. **Spin LOR-3A at 500rpm (2seconds) followed by 3000rpm (30 seconds),** using program B of our spinner.
2. **Bake** on hotplate **at 170C (175°C in our plate) for 5 minutes.**
3. **Spin S1813 at 500rpm (2seconds) followed by 5000rpm (30 seconds),** using program A of our spinner.
4. **Bake** on hotplate **at 115C (120°C in our plate) for 2 minutes.**
5. **UV-expose for 7 seconds.**

6. **Develop in CD26 for 45 seconds.** Rinse in DI-water and blow dry.
7. Hotplate **bake at 125C (130°C in our plate) for 5 minutes**
8. Undercut **etch in CD26 for 1 minute (do not move the sample while developing in this step)**. Rinse in DI-water and blow dry.

Depositing the contacts:

1. (Optional. This step is not required if the sample has been prepared right before deposition) Removal of the oxide layer on the surface of the sample: **Deep the sample in DI H2O:Ammonium Hydroxide (4:1) for 5 seconds** immediately before putting the sample inside the UHV chamber.
2. Deep the sample in DI water for 15 seconds.
3. Blow dry.
4. Deposit the contacts: Use small deposition rates for this.
    - 5nm thickness Pt (Bulk density 21.37, z-ratio 0.245) or Cr
    - 100nm thickness Au/Ge 12% alloy (bulk density 14.68, acoustic impedance 22.45)
    - Optional (50-100nm thickness Au on top)
    - Use the tooling factor for the thermal and e-beam evaporators.
5. Lift-off the metal outside the contacts: deep the sample in acetone (if single-layer process is been used) or in remover PG (if double-layer process) and ultrasound for short periods of time (1 second each period or less) until having all the photoresist out. Ultrasound bath is a very aggressive method that can peal off the contacts pads, so its use at this stage must be minimized.
6. Clean the sample with acetone/isopropanol/ethanol WITHOUT ultrasounds.

H. ANNEALING THE CONTACTS

1. Check what are the approximate values of the current needed for each of the temperatures of the annealing procedure before placing the sample into the annealing box.
2. If possible, place a microscope on top of the box to check how the metal melts during the annealing process.
3. Place the sample on the grid inside the Rapid Thermal Annealing box close to the reference sample. The thermometer must be touching the surface of a similar thickness and composition piece of wafer to have similar reading of the temperature of the surface of the sample.
4. A continuous flow of He needs to be generated inside the box. Tune the flow with the needle valve.
5. To anneal the contacts follow the next procedure:
    - 110°C for 60 seconds
    - 250°C for 10 seconds
    - 410°C for 20-30 seconds

I. CHECKING THE HALL SENSOR AT LOW TEMPERATURE

1. Make contacts to the leads of the sensor with the wire bonder.
2. Measure the resistance at RT using the low current voltmeter in MAP lab. Never use a Fluke (yellow) voltmeter in doing this, high currents will break the sensor.
3. RT resistances for a 50μm × 50μm sensor should be in the order of 20-50kΩ.
4. Helium temperature (4.2K) resistance should be in the order of 1-10kΩ.
5. Apply a low frequency (100Hz-100kHz) ac current of 1-5μA between two of the leads of the sensor.
6. Measure the Hall voltage between the opposite set of leads of the sensor. Use a Lock-in amplifier at the same frequency of the current to increase sensitivity.
7. Sweep a magnetic field from negative 1-2T to positive 1-2T. The field needs to be applied perpendicular to the plane of the sensor.
8. This can be done in the homemade closed cycle cryostat at 10 K using Peale's electromagnet instead of liquid Helium.
9. The Hall voltage is proportional to the field magnitude:

$$V_H = \frac{R_H}{t} GIB$$

10. The Hall coefficient is calculated as follows:

$$R_H = \frac{V_H}{IB} \left[= \frac{V}{AT} = \frac{\Omega}{T} \right]$$

A typical value for our sensors is 1,800 Ω/T.

11. To calculate the density of carriers use:

$$n = \frac{1}{eR_H} \left[= \frac{T}{\Omega C} = \frac{1}{m^2} \right]$$

A typical value for our material is $2-4\times10^{11}$ cm^{-2}

REFERENCES

[1] Y. Nakamura, Yu. A. Pashkin, and J. S. Tsai, *Coherent Control of Macroscopic Quantum States in a Single-Cooper-Pair Box*. Nature **398**, 786 (1999)

[2] T. Hayashi, T. Fujisawa, H. D. Cheong, Y. H. Jeong, and Y. Hirayama, *Coherent Manipulation of Electronic States in a Double Quantum Dot*, Phys. Rev. Lett. **91**, 226804 (2003).

[3] M. N. Leuenberger and D. Loss, *Quantum Computing in Molecular Magnets*, Nature **410**, 789 (2001); J. Tejada, E. M. Chudnovsky, E. del Barco, and J. M. Hernandez, *Magnetic Qubits as Hardware for Quantum Computers*, Nanotechnology **12**, 181 (2001).

[4] W. Wernsdorfer and R. Sessoli, *Quantum Phase Interference and Parity Effects in Magnetic Molecular Clusters*, Science **284**, 133 (1999).

[5] E. del Barco, A. D. Kent, E. M. Lumberger, D. N. Hendrikson, and G. Christou, *Symmetry of Magnetic Quantum Tunneling in Single Molecule Magnet Mn_{12}-acetate*, Phys. Rev. Lett. **91**, 047203 (2003).

[6] E. del Barco, A. D. Kent, S. Hill, J. M. North, N. S. Dalal, E. M. Rumberger, D. N. Hendrickson, N. Chakov, and G. Christou, *Magnetic Quantum Tunneling in the Single-Molecule Magnet Mn_{12}-acetate*, J. Low. Temp. Phys. **140**, 119 (2005).

[7] J. R. Friedman, M. P. Sarachik, J. Tejada, and R. Ziolo, *Macroscopic Measurement of Resonant Magnetization Tunneling in High-Spin Molecules*, Phys. Rev. Lett. **76**, 3830 (1996); J. M. Hernández, X. X. Zhang, F. Luis, J. Bartolomé, J. Tejada, and R. Ziolo, *Field Tuning of Thermally Activated Magnetic Quantum Tunneling in Mn_{12}-Ac Molecules*, Europhys. Lett. **35**, 301 (1996); L. Thomas, F. Lionti, R. Ballou, D. Gatteschi, R. Sessoli, and B. Barbara, *Macroscopic Quantum Tunneling of Magnetization in a Single Crystal of Nanomagnets*, Nature **383**, 145 (1996).

[8] E. M. Chudnovsky and J. Tejada, *Macroscopic Quantum Tunneling of Magnetic Moment*. Cambridge University Press, Cambridge, England, (1998); C. Sangregorio, T. Ohm, C. Paulsen, R. Sessoli, and D. Gatteschi, *Quantum Tunneling of the Magnetization in an Iron Cluster Nanomagnet*, Phys. Rev. Lett. **78**, 4645 (1997); A. L. Barra, D. Gatteschi, and R. Sessoli, *High-Frequency EPR Spectra of a Molecular Nanomagnet: Understanding Quantum Tunneling of the Magnetization*, Phys. Rev. B**56**, 8192 (1997); S. Hill, J. A. A. J. Perenboom, N. S. Dalal, T. Hathaway, T. Stalcup, and J. S. Brooks, *Highs-Sensitivity Electron Paramagnetic Resonance of Mn_{12}-acetate*, Phys. Rev. Lett. **80**, 2453 (1998); I. Mirebeau, M. Hennion, H. Casalta, H. Andres, H. U. Güdel, A. V. Irodova, and A. Caneschi, *Low-Energy Magnetic Axcitations of the Mn_{12}-acetate Spin Cluster Observed by Neutron Scattering*, Phys. Rev. Lett. **83**, 628 (1999); F. Luis, F. Mettes, L. J. De Jongh, J, Tejada, and D. Gatteshi, *Observation of Quantum Coherence in Mesoscopic Molecular Magnets*, Phys. Rev. Lett. **85**, 4377 (2000); R. Caciuffo, G. Amoretti, A. Murani, R. Sessoli, A. Caneschi, and D. Gatteschi, *Neutron Spectroscopy for the Magnetic*

Anisotropy of Molecular Clusters, Phys.Rev. Lett. **81**, 4744 (1998); F. Luis, J. Bartolomé, and J. Fernandez, *Resonant Magnetic Quantum Tunneling through Thermally Activated states*, Phys. Rev. B**57**, 505 (1998); J. F. Fernandez, F. Luis, and J. Bartolomé, *Time Dependent Specific Heat of a Magnetic Quantum Tunneling System*, Phys. Rev. Lett. 80, 5659 (1998); E. del Barco, N. Vernier, J. M. Hernandez, J. Tejada, E. M. Chudnovsky, E. Molins, and G. Bellesa, *Mesoscopic Quantum Coherence in High-Spin Molecules*, Europhys. Lett. **47**, 722 (1999); E. del Barco, J. M. Hernandez, J. Tejada, N. Biskup, R. Achey, I Rutel, N. Dalal, and J. Brooks, *High Frequency Resonance Experiments on Fe-8 Molecular Clusters*, Phys. Rev. B**62**, 3018 (2000).

[9] O. Waldmann, C. Dobe, H. Mutka, A. Furrer, and H. U. Gudel, *Phys. Rev. Lett.* **95**, 057202 (2005).

[10] P. King, T. C. Stamatatos, K. A. Abboud and G. Christou, *Reversible size modification of iron and gallium molecular wheels: A Ga_{10} "gallic wheel" and large Ga_{18} and Fe_{18} wheels.* Angew. Chem. Int. Ed. **45**, 7379-7383 (2006).

[11] F. Troiani *et al.*, *Molecular engineering of antiferromagnetic rings for quantum computation.* Phys. Rev. Lett. **94**, 207208 (2005).

[12] A. Ardavan, O. Rival, J. J. L. Morton, S. J. Blundell, A. M. Tyryshkin, G. A. Timco and R. E. P. Winpenny, *Will spin-relaxation times in molecular magnets permit quantum information processing?* Phys. Rev. Lett. **98**, 057201 (2007).

[13] S. Bertaina, S. Gambarelli, T. Mitra, B. Tsukerblat, A. Müller, and B. Barbara, Nature London **453**, 203 2008.

[14] C. Schlegel, J. van Slageren, M. Manoli, E. K. Brechin, and M.Dressel, Phys. Rev. Lett. **101**, 147203 2008.

[15] S. Takahashi, J. van Tol, C. C. Beedle, D. N. Hendrickson, L.-C.Brunel, and M. S. Sherwin, arXiv:0810.1254 unpublished.

[16] S. Takahashi, R. Hanson, J. van Tol, M. S. Sherwin, and D. D. Awschalom, Phys. Rev. Lett. **101**, 047601 2008.

[17] P. King, T. C. Stamatatos, K. A. Abboud, and G. Christou, Angew. Chem, Int. Ed. **45**, 7379 2006.

[18] G. L. Abbati, L.C. Brunel, H. Casalta, A. Cornia, A. C. Fabretti, D. Gatteschi, A. K. Hassan, A. G. M. Jansen, A. L. Maniero, L. Pardi, C. Paulsen, and U. Segre, Chem. - Eur. J. 7, 1796 2001.

[19] E. M. Chudnovsky, Phys. Rev. Lett. **92**, 120405 2004.

[20] P. L. Feng *et al.*, *Inorg. Chem.* **47**, 8610-8612 (2008).

[21] P. L. Feng *et al.*, *Inorg. Chem.* **48**, 3480-3492 (2009).

[22] R. Inglis *et al.*, *Chem. Eur. J.* **14**, 9117-9121 (2008).

[23] R. Inglis *et al.*, *Chem. Commun.* 5924-5926 (2008).

[24] S. G. Sreerama and S. Pal, *Inorg. Chem.* **41**, 4843 (2002).

[25] C-I. Yang, W. Wernsdorfer, K-H. Cheng, M. Nakano, G-H. Lee and H-L. Tsai, *Inorg. Chem.* **47**, 10184 (2008).

[26] T. C. Stamatatos *et al.*, *J. Am. Chem. Soc.* **129**, 9484 (2007).

[27] W. Wernsdorfer and R. Sessoli, *Science* **284**, 133-135 (1999).

[28] K. M. Mertes *et al.*, *Phys. Rev. Lett.* **87**, 227205 (2001).

[29] A. Cornia, R. Sessoli, L. Sorace, D. Gatteschi, A. L. Barra and C. Daiguebonne, *Phys. Rev. Lett.* **89**, 257201 (2002).

[30] S. Takahashi, R. S. Edwards, J. M. North, S. Hill and N. S. Dalal, *Phys. Rev. B* **70**, 094429 (2004).

[31] E. del Barco *et al.*, *J. Low. Temp. Phys.* **140**, 119-174 (2005).

[32] M. Schechter and P. C. E. Stamp, Phys. Rev. Lett. 95, 267208 (2005).

[33] J. Lawrence, E.-C. Yang, R. Edwards, M. M. Olmstead, C. Ramsey, N. S. Dalal, P.K. Gantzel, S. Hill, D. N. Hendrickson, *Inorg. Chem.* **47**, 1965-1974 (2008).

[34] R. Inglis *et al.*, *Chem. Eur. J.* **14**, 9117-9121 (2008).

[35] R. Inglis *et al.*, *Chem. Commun.* 5924-5926 (2008).

[36] S. G. Sreerama and S. Pal, *Inorg. Chem.* **41**, 4843 (2002).

[37] C-I. Yang, W. Wernsdorfer, K-H. Cheng, M. Nakano, G-H. Lee and H-L. Tsai, *Inorg. Chem.* **47**, 10184 (2008).

[38] T. C. Stamatatos *et al.*, *J. Am. Chem. Soc.* **129**, 9484 (2007).

[39] A. D. Kent *et al. J. Appl. Phys.* **76**, 6656 (1994).

[40] L. Bokacheva, A. D. Kent and M. A. Walters, *Phys. Rev. Lett.* **85**, 4803-4806 (2000).

[41] Wilson, J. Lawrence, E-C. Yang, M. Nakano, D. N. Hendrickson and S. Hill, *Phys. Rev. B* **74**, 140403(R) (2006).

[42] Mantel *et al.*, *J. Am. Chem. Soc.* **125**, 12337-12344 (2003).

[43] M. A. Reed, C. Zhou, C. J. Muller, T. P. Burgin, and J. M. Tour, *Conductance of a Molecular Junction*, Science **278**, 252 (1997).

[44] M. N. Leuenberger and E. R. Mucciolo, *Berry-Phase Oscillations of the Kondo Effect in Single-Molecule Magnets*, Phys. Rev. Lett. **97**, 126601 (2006).

[45] G. Gonzalez and M. N. Leuenberger, Berry-Phase Blockade in Single-Molecule Magnets, preprint (cond-mat/0610653).

[46] B. J. Kim, B. J. Suh, S. Yoon, S. Phark, Z. G. Khim, J. Kim, J. M. Lim, and Y. Do, J. Korean Phys. Soc. **45**, 1593 _2004_; S. Phark, Z. G. Khim, B. J. Kim, B. J. Suh, and S. Yoon, Jpn. J. Appl. Phys., Part 1 **43**, 8273 (2004).

[47] J. Park, Ph.D. thesis, University of California-Berkeley, (2003).

[48] C. Durkan, M. A. Schneider, and M. E. Weilland, J. Appl. Phys. **86**, 1280 (1999).

[49] T. Lis, *Preparation, Structure, and Magnetic Properties of a Dodecanuclear Mixed Valence Manganese Carboxylate*, Acta Crystallogr. Sect. B, **B36**, 2042 (1980).

[50] R. Sessoli, H.-L. Tsai, A. R. Schake, S. Wang, J. B. Vincent, K. Folting, D. Gatteschi, G. Christou, and D. N. Hendrickson, *High-Spin Molecules: [Mn12O12(O2CR)16(H2O)4]*, J. Am. Chem. Soc. **115**, 1804-1816 (1993).

[51] H. J. Eppley, H.-L. Tsai, N. de Vries, K. Folting, G. Christou, and D. N. Hendrickson, *High Spin Molecules: Unusual Magnetic Susceptibility Relaxation Effects in [Mn12O12(O2CEt)16(H2O)3] (S=9) and the One-Electron Reduction Product (PPh4)[Mn12O12(O2CEt)16(H2O)4] (S=19/2)*, J. Am. Chem. Soc. **117**, 301 (1995).

[52] M. Soler, W. Wernsdorfer, K. A. Abboud, J. C. Huffman, E. R. Davidson, D. N. Hendrickson, and G. Christou, *Single-Molecule Magnets: Two-Electron Reduced Version of a Mn12 Complex, and Environmental Influences on the Magnetization Relaxation of (PPh4)2[Mn12O12(O2C-CHCl2)16(H2O)4]*, J. Am. Chem. Soc. **135**, 3576 (2003).

[53] M. W. Wemple, D. M. Adams, K. S. Hagen, K. Folting, D. N. Hendrickson, and G. Christou. *Site-Specific Ligand Variation in Manganese-Oxide Cubane Complexes, and Unusual Magnetic Relaxation Effects in [Mn4O3X(OAc)3(dbm)3] (X = N3, OCN)*. J. Chem. Soc., Chem. Commun. 1591 (1995).

[54] S. Wang, H.-L. Tsai, K. Folting, W. E. Streib, D. N. Hendrickson, and G. Christou. *Modeling the Photosynthetic Water Oxidation Center: Chloride/Bromide Incorporation and Reversible Redox Processes in the Complexes Mn4O3X(OAc)3(dbm)3 (X = Cl, Br)*, Inorg. Chem. **35**, 7578 (1996).

[55] S. M. J. Aubin, M. W. Wemple, D. M. Adams, H.-L.Tsai, G. Christou, and D. N. Hendrickson, *Distorted MnIVMnIII 3 Cubane Complexes as Single-Molecule Magnets*, J. Am. Chem. Soc. **118**, 7746 (1996).

[56] H. B. Heersche, Z. de Groot, J. A. Folk, H. S. J. van der Zant, C. Romeike, M. R. Wegewijs, L. Zobbi, D. Barreca, E. Tondello, and A. Cornia, *Electron Transport through Single Mn12 Molecular Magnets*, Phys. Rev. Lett. **96**, 206801 (2006).

[57] M.-H. Jo, J. E. Grose, K. Baheti, M. M. Deshmukh, J. J. Sokol, E. M. Rumberger, D. N. Hendrickson, J. R. Long, H. Park, and D. C. Ralph, *Signatures of Molecular Magnetism in Single-Molecule Transport Spectroscopy*, Nano Lett. **6**, 2014-2020 (2006).

CPSIA information can be obtained
at www.ICGtesting.com
Printed in the USA
LVIW021427021012
301197LV00005B